Lecture Notes in Computational Science and Engineering

89

Editors:

Timothy J. Barth
Michael Griebel
David E. Keyes
Risto M. Nieminen
Dirk Roose
Tamar Schlick

Michael Griebel   •   Marc Alexander Schweitzer

Editors

Meshfree Methods for Partial Differential Equations VI

 Springer

Editors
Michael Griebel
Institut für Numerische Simulation
Universität Bonn
Bonn
Germany

Marc Alexander Schweitzer
Institut für Parallele und Verteilte Systeme
Universität Stuttgart
Stuttgart
Germany

ISSN 1439-7358
ISBN 978-3-642-42977-4   ISBN 978-3-642-32979-1 (eBook)
DOI 10.1007/978-3-642-32979-1
Springer Heidelberg New York Dordrecht London

Mathematics Subject Classification (2010): 65N99, 64M99, 65M12, 65Y99

Springer is part of Springer Science+Business Media (www.springer.com)

Preface

The Sixth International Workshop on *Meshfree Methods for Partial Differential Equations* was held from October 4–6, 2011, in Bonn, Germany. One of the major goals of this workshop series was to bring together European, American and Asian researchers working in this exciting field of interdisciplinary research on a regular basis. To this end, Ivo Babuška, Ted Belytschko, Jiun-Shyan Chen, Michael Griebel, Wing Kam Liu, Marc Alexander Schweitzer and Harry Yserentant invited scientists from all over the world to Bonn to strengthen the mathematical understanding and analysis of meshfree discretizations, but also to promote the exchange of ideas on their implementation and application.

The workshop was again hosted by the Institut für Numerische Simulation at the Rheinische Friedrich-Wilhelms-Universität Bonn with the financial support of the Sonderforschungsbereich 611 *Singular Phenomena and Scaling in Mathematical Models*. Moreover, we would like to thank Christian Rieger who carried most of the load as the local organizer of this workshop.

This volume of LNCSE now comprises selected contributions of attendees of the workshop. Their content ranges from applied mathematics to physics and engineering which clearly indicates the maturity meshfree methods have reached in recent years. They are becoming more and more mainstream in many fields due to their flexibility and wide applicability.

Bonn and Stuttgart, Germany

Michael Griebel
Marc Alexander Schweitzer

Contents

ESPResSo 3.1: Molecular Dynamics Software for Coarse-Grained Models

Axel Arnold, Olaf Lenz, Stefan Kesselheim, Rudolf Weeber,
Florian Fahrenberger, Dominic Roehm, Peter Košovan, and Christian Holm

Abstract ESPReSSo is a package for Molecular Dynamics (MD) simulations of coarse-grained models. We present the most recent version 3.1 of our software, highlighting some recent algorithmic extensions to version 1.0 presented in a previous paper (Limbach et al. Comput Phys Commun 174:704–727, 2006). A major strength of our package is the multitude of implemented methods for calculating Coulomb and dipolar interactions in periodic and partially periodic geometries. Here we present some more recent additions which include methods for systems with dielectric contrasts that frequently occur in coarse-grained models of charged systems with implicit water models, and an alternative, completely local electrostatic solver that is based on the electrodynamic equations. We also describe our approach to rigid body dynamics that uses MD particles with fixed relative positions. ESPReSSo now gained the ability to add bonds during the integration, which allows to study e.g. agglomeration. For hydrodynamic interactions, a thermalized lattice Boltzmann solver has been built into ESPReSSo, which can be coupled to the MD particles. This computationally expensive algorithm can be greatly accelerated by using Graphics Processing Units. For the analysis of time series spanning many orders of magnitude in time scales, we implemented a hierarchical generic correlation algorithm for user-configurable observables.

Keywords Molecular dynamics • Coarse-graining • Lattice-Boltzmann

A. Arnold (✉) · O. Lenz · S. Kesselheim · R. Weeber · F. Fahrenberger · D. Roehm · P. Košovan · C. Holm
Institute for Computational Physics, Universität Stuttgart, Allmandring 3, 70569 Stuttgart, Germany
e-mail: arnolda@icp.uni-stuttgart.de

M. Griebel and M.A. Schweitzer (eds.), *Meshfree Methods for Partial Differential Equations VI*, Lecture Notes in Computational Science and Engineering 89, DOI 10.1007/978-3-642-32979-1_1, © Springer-Verlag Berlin Heidelberg 2013

1 Introduction

Nowadays, computer simulations are a well established tool in theoretical physics. Here we intend to give an introduction to our Molecular Dynamics package ESPRESSO [6, 26]; ESPRESSO is an acronym for **E**xtensible **S**imulation **P**ackage for **Res**earch on **So**ft matter systems.

The term *soft matter*, or *complex fluids*, as they are called in the American literature, describes a large class of materials, such as polymers, colloids, liquid crystals, glasses, hydrogels and dipolar fluids; familiar examples of such materials are glues, paints, soaps or baby diapers. Also most biological materials are soft matter – DNA, membranes, filaments and other proteins belong to this class. Soft matter research has experienced an increased interest in the last two decades due to its potentially high usefulness in many areas such as technology, biophysics, and nanoscience.

Many properties of soft matter materials emerge on the molecular rather than the atomistic level: the elasticity of rubber is the result of entropy of the long polymer molecules, and the superabsorbing materials used in modern diapers store water inside a polyelectrolyte network. To reproduce these physical effects on the atomistic level in computer simulations, one would have to incorporate many millions of atoms and simulate them for time scales of up to seconds in some cases, which is not possible even with the most powerful modern computers. However, in many cases a much simpler description of the material is sufficient. Polymers such as polyelectrolytes or rubber often can be modeled by simple *bead-spring models*, i.e. (charged) spheres connected by springs, where each of the spheres represents a whole group of atoms, sometimes a complete monomer or even larger compounds. Although these models hide most of the chemical properties, they are quite successful in the description of polymers and other soft matter systems. The process of removing degrees of freedom (here the atomistic description) from a system to obtain a simpler model is called *coarse-graining*.

Computer simulations of such coarse-grained models sometimes still require the incorporation of several thousands of beads and springs, and many time steps (in case of MD) for data acquisition, and therefore one still is in the need of an efficient simulation software. Furthermore, the generation of sufficient amounts of uncorrelated data might still require the usage of non-standard algorithms (i.e. Gibbs ensemble, expanded ensemble techniques, AdResS [22] just to name a few) which in some cases even have been developed especially for a specific model. Therefore it is necessary that the simulation software is much more flexible than standard atomistic simulation packages (as for example GROMACS [18][1] or NAMD [35][2]). ESPRESSO was designed and implemented specifically to address

---

[1]http://www.gromacs.org
[2]http://www.ks.uiuc.edu/Research/namd/

the requirements of simulating such coarse-grained models, and it has a number of unique characteristics that distinguish it from any other simulation package that we know of.

One should be aware that the flexibility of ESPRESSO also costs some performance: compared to highly optimized MD programs such as GROMACS, ESPRESSO is slower by a factor of about 2. However, one should keep in mind that most of the problems, that we designed ESPRESSO for, can normally not be treated without massive changes to the simulation engines of these fast codes.

ESPRESSO is not a self-contained package but relies on other open source packages. Most prominent is the use of the Tcl scripting language interpreter for the simulation control, which is required to run ESPRESSO. Other packages are optional. If ESPRESSO is to be executed in parallel, an implementation of the MPI standard [32] is required. The P^{3}M algorithm for electrostatic interactions relies on the FFTW package.[3] The development process is supported heavily by the use of the git version control system,[4] which allows several developers to work simultaneously on the code, LATEX and doxygen[5] for the documentation, and the GNU Autotools[6] for compilation.

In the following Sect. 2, we will detail the unique characteristics of the ESPRESSO simulation software, while in Sect. 3 we will give an overview of the various algorithms and methods implemented in the software. The subsequent Sects. 4–8 will describe the recent algorithmic developments in ESPRESSO. We conclude with an outlook on future plans for our software in Sect. 9.

2 Characteristics

Optimized for coarse-grained models ESPRESSO was explicitly designed for simulating coarse-grained models. On the one hand, this means that the software has a few characteristics that are otherwise uncommon to most simulation software, such as the fact that the software is *controlled by a scripting language* or that it is *extensible* (see below). On the other hand, the software implements a number of special methods that only make sense in the context of coarse-grained models, such as the possibilities to create rigid bodies (see Sect. 5), to dynamically generate new bonds during the simulation (see Sect. 6), or to couple the many-particle simulation to a Lattice Boltzmann (LB) fluid (see Sect. 7), to name just a few.

---

[3]http://fftw.org
[4]http://git-scm.com
[5]http://doxygen.org
[6]http://ftp.gnu.org/gnu/automake/

Also note that the software does not enforce any specific length, energy or time unit system, as such units are often inappropriate for coarse-grained models. Instead, the values can be given in any consistent unit system.

Controlled by a scripting language   The software uses the scripting language Tcl[7] to control the simulations. However, as the Tcl language is mostly outdated nowadays, we are currently working on a new interface to the scripting language Python[8] for one of the next versions of ESPRESSO.

The simulation control script determines all simulation parameters such as the number and type of particles, the type of interactions between these particles, and how the system is propagated; most of the parameters can be changed even during runtime of the simulation. By that, one can perform highly complex simulation procedures, such as adapting the interaction parameters to the current configuration during the simulation, applying or removing spatial constraints, or even complex schemes like parallel tempering or hybrid Monte Carlo. This flexibility is unmatched by any other simulation package that we know of.

Extensible   Users can read and modify the code to meet their own needs. Throughout the source code of ESPRESSO, readability is preferred over code optimizations. This allows users to extend the code with more ease. Furthermore, we have defined a number of interfaces that allow to implement extensions to the core code (e.g. new potentials, new thermostats or new analysis routines). The extensibility allows researchers to easily incorporate new methods and algorithms into the software, so that ESPRESSO can be used both as a production platform to generate simulation data as well as a research platform to test new algorithms.

Free and open source   ESPRESSO is an open-source program that is published under the GNU public license. Both the source code of release versions of the software as well as the bleeding-edge development code, are available through our web page[9] and the GNU Savannah project page.[10]

Parallelized   ESPRESSO is parallelized, allowing for simulations of millions of particles on hundreds of CPUs. ESPRESSO scales well, it can achieve an efficiency of about 70 % on 512 Power4+ processors, and even better values on the more recent BlueGene/P or Cray XT6 systems. Since it contains some of the fastest currently available simulation algorithms, it also scales well with the number of particles, allowing for the simulation of large systems.

Portable   The code kernel is written in simple ANSI C, therefore it can be compiled on a wide variety of hardware platforms like desktop workstations, convenience clusters and high performance supercomputers based on POSIX operating systems (for example all variants of Unix that we know of, including GNU/Linux).

---

[7]http://www.tcl.tk

[8]http://www.python.org

[9]http://espressomd.org

[10]http://savannah.nongnu.org/projects/espressomd/

3 Methods and Algorithms

In the following, we will give an overview of the algorithms and methods that ESPRESSO provides. Some of these are standard algorithms, like they are described in, e.g., the book Understanding Molecular Simulation [15].

Ensembles The software can perform MD simulations in a number of different physical ensembles, using the Velocity-Verlet integration scheme. Besides the microcanonical (NVE) ensemble, it is possible to simulate the canonical (NVT) ensemble via the Langevin thermostat and the NPT (constant pressure) ensemble via a barostat algorithm by Kolb and Dünweg [24]. The constant chemical potential ensemble (μVT) can be realized by hybrid Molecular Dynamics/Monte Carlo on the script language level.

Nonbonded potentials For nonbonded interactions between different particle species, a number of different potentials are implemented, for example the Lennard-Jones, Morse and Buckingham potentials. In addition, it is possible to use tabulated potentials of arbitrary form.

Bonded potentials ESPRESSO provides a number of interactions between two or more specific particles, including the FENE and harmonic bond potentials, bond angle and dihedral interactions. As in the nonbonded case, potentials in arbitrary form can be specified via tables.

Anisotropic interactions Besides the standard integrator, the software also has a quaternion integrator [30] where particles have an orientation represented by a quaternion and rotational degrees of freedom. Therefore, particles cannot only represent isotropic spheres, but also objects that interact via anisotropic interactions such as Gay-Berne ellipsoids [16].

Electro- and Magnetostatics The package implements a number of fast parallel state-of-the-art algorithms for electrostatic and magnetostatic interactions that can handle full or partial periodicity, and that even allow to treat systems with dielectric contrast. The methods are detailed in Sect. 4.

Constraints ESPRESSO has the ability to fix some or all coordinates of a particle, or to apply an arbitrary external force on each particle. In addition, various spatial constraints, such as spherical or cubic compartments, can be used. These constraints interact with the particles by any nonbonded interaction.

Rigid bodies Since version 3.0, it is possible to form rigid bodies out of several particles. This feature allows for arbitrarily complex extended objects in the model, and is described in Sect. 5.

Dynamic bonding Agglomeration can be modeled by dynamically adding bonds during the simulation when particles collide. This special feature is described in Sect. 6.

Hydrodynamics Hydrodynamic interactions can be modeled via a thermal Lattice Boltzmann method (see Sect. 7) that can be coupled to the particles. To accelerate the algorithm, it is possible to run it on a graphics processor (GPU). Alternatively, particles can use Dissipative Particle Dynamics (DPD) [14, 40] to simulate hydrodynamic effects.

Analysis   All analysis routines are available in the simulation engine, allowing
for both online analysis (during the simulation) as well as offline analysis.
ESPRESSO can measure various observables of the system, such as the energy
and (isotropic) pressure, or the forces acting on particles or spatial constraints.
There are routines to determine particle distributions and structure factors, and
some polymer-specific measures such as the end-to-end distance or the radius
of gyration. For visualization, ESPRESSO can output files that can be read by
visualization software such as VMD[11] [20]. It is simple for users to add their own
observables. To allow for measuring correlations in timeseries (such as the mean-
square displacement of particles), ESPRESSO contains a generic implementation
of *correlators*, which is detailed in Sect. 8.

AdResS   ESPRESSO contains an implementation of the **Ad**aptive **Res**olution
Scheme (AdResS) that allows to simulate systems that contain areas with
different levels of coarse-graining [22].

4 Advanced Electrostatics

Coulomb interactions cannot be calculated by the same straightforward neighborhood strategies as short-ranged interactions, due to the slow decay of the Coulomb
potential. Especially in conjunction with periodic (or partially periodic) boundary
conditions, one needs special algorithms that are adapted to the particular periodic
structure. For systems with three periodic dimensions, which is the most common
case as it is used for investigating bulk systems, ESPRESSO offers the P^3M [10, 11]
and MEMD algorithms [28, 34]. P^3M is also able to treat dipolar systems, such
as ferrofluids [9]. For systems with only two periodic dimensions and a third
dimension with finite extension, which are commonly used to investigate thin
films or surfaces, ESPRESSO features the MMM2D [3] and ELC methods [5],
and for rod-like systems, such as nanotubes or pores, only one periodic and two
nonperiodic dimensions are required. For these systems, ESPRESSO offers the
MMM1D algorithm [4].

Since the last publication on ESPRESSO [26], the electrostatics algorithms have
been considerably enhanced. The P^3M implementation allows now to use arbitrary
simulation box dimensions, i.e. not just cubic boxes. Still, our implementation
features full error control and automatic parameter tuning, as described by Deserno
et al. [11]. Other extensions that are to our knowledge unique to ESPRESSO will
be described in this section. First, we will discuss extensions to the MMM2D
and ELC methods, which allow to take into account the presence of planar
dielectric interfaces. Next, we will introduce the ICC⋆ method, which allows to

---

[11]http://www.ks.uiuc.edu/Research/vmd/

take into account arbitrarily shaped dielectric interfaces. Finally, we will discuss the MEMD electrostatics algorithm, a local approach to electrostatics, based on the electrodynamic equations that use an appropriately adapted speed of light.

4.1 Dielectric Contrasts

In coarse-grained systems with only partially periodic boundaries and implicit water models, where the water is modeled as a dielectric continuum with a relative dielectric constant, the dielectric contrasts between the embedding medium and the outside can be quite considerable. For example, the relative dielectric constant for bulk water at room temperature is 80, whereas it has a value of ≈ 1 in the surrounding air. When studying ions in front of a metallic electrode, the latter even has an infinite dielectric constant. Due to the different dielectric media, polarization occurs, which has a non-negligible influence on the charges in these systems. For example, Messina [31] has shown that image charges due to polarization may lead to considerable reduction in the degree of polyelectrolyte adsorption onto an oppositely charged surface and by that inhibit charge inversion of the substrate.

In computational studies, these polarization effects need to be taken into account. At present, ESPResSo supports this by extensions to the ELC and MMM2D algorithms, or by the novel ICC⋆ algorithm, as described below. The first extensions reach a higher precision and are faster, but less flexible than the ICC⋆ method and can handle only two planar, parallel dielectric interfaces that can have, however, arbitrary dielectric contrast. These need to be parallel to the two periodic dimensions, which is sufficient to model both the two interfaces surrounding a thin film, or the single dielectric jump at a wall. The dielectric boundary conditions are taken into account by the method of image charges [39]. We assume a system, where the charges are embedded in a medium characterized by a dielectric constant ε_m which is confined from above by a medium of dielectric constant ε_t and from below by a medium of dielectric constant ε_b, compare Fig. 1. In case only a single interface should be considered, this can be achieved by choosing $\varepsilon_t = \varepsilon_m$ or $\varepsilon_b = \varepsilon_m$.

In the general case of two dielectric boundaries, an infinite number of image charges arises for each single physical charge, compare Fig. 1. For example, a charge q at a position z gives rise to an image charge of $\Delta_b q$ at $-z$ in the lower dielectric layer and an image charge of $\Delta_t q$ at $(2l_z - z)$ in the upper dielectric layer. The image charge $\Delta_b q$ gives rise to another image charge $\Delta_t \Delta_b q$ in the top dielectric domain. Similarly $\Delta_t q$ gives rise to an image charge $\Delta_b \Delta_t q$ in the bottom dielectric domain, and so on. Here, the prefactors Δ_b and Δ_t are defined as

$$\Delta_b = \frac{\varepsilon_m - \varepsilon_b}{\varepsilon_m + \varepsilon_b}, \quad \text{and} \quad \Delta_t = \frac{\varepsilon_m - \varepsilon_t}{\varepsilon_m + \varepsilon_t}.$$

These polarization image charges constitute simple, but infinite geometric series that can be taken into account analytically by the far formula of the MMM2D method [3]. This formula gives a simple expression for the interaction energy

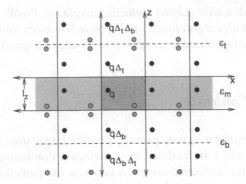

Fig. 1 Image charges generated by dielectric contrasts. The dielectric interfaces are characterized by the planar ε_t–ε_m and ε_m–ε_b boundaries. Polarization leads to an infinite number of image charges along the z-direction due to multiple reflections. Additionally, periodic boundary conditions apply in x and y directions. The *dotted lines* are only provided to visualize the positioning of the image charges

or force between a charge and a periodic array of charges. Since this formula is also at the heart of the ELC method, both MMM2D and ELC can take these additional image charges into account by prefactors to the already present far formula implementations. These prefactors can be conveniently derived from analytic summations of the geometric sums [42, 43].

MMM2D and ELC allow to specify Δ_b and Δ_t directly instead of specifying the three dielectric constants. By choosing these values as ± 1, one can obtain effectively infinite dielectric constants inside or outside. Finally, note that the method of image charges describes only the correct electrostatic potential inside the dielectric slab; therefore, particles need to be constrained to the area between the two dielectric surfaces.

4.2 MEMD

The MEMD algorithm (**M**axwell **E**quations **M**olecular **D**ynamics) is a rather uncommon way to compute electrostatic interactions, based on the full electrodynamic Maxwell equations. Of course, simulating with actual realistic electrodynamics is by far too time consuming, but it can be made more effective with some adaptions. Most notably, the speed of light needs to be reduced to become comparable to the other degrees of freedom, that is, the speed of the atoms and molecules in the simulation.

Algorithm

The algorithm starts with a regular discretization of the Maxwell equations. To overcome the problem of the large value for the light speed, A. Maggs has shown [28] that the speed of light can be tuned to very small values, as long as it stays well above the speed of the charges. In this case, the full dynamics of the theory (i.e. for the magnetic fields) can be unrealistic while the electrostatic limit is still valid. In MD simulations, the speed of light can be chosen by about an order of magnitude larger than typical particle velocities, which however are small compared to the actual speed of light.

Another very important adaptation is, that this method actually consists of two different combined methods: initially, a solution of the Poisson equation for the system is computed with a numerical relaxation scheme. Afterwards the correct solution can be obtained by only applying temporal updates every time step. The time derivative and some physical arguments as laid out in Ref. [34] lead to the following very simple constraint that can be applied to the propagation of the system:

$$\frac{\partial}{\partial t}\mathbf{D} + \mathbf{j} - \frac{1}{c^2}\nabla \times \mathbf{B} = 0$$

with the electric field $\mathbf{D}$, the current $\mathbf{j}$ and a magnetic field component $\mathbf{B}$ that is propagating in dual space.

This constraint is now ensured by interpolating the charges (or to be more specific, the electric currents) onto a regular lattice. Then the magnetic fields that are created from the current are propagated on the lattice. From these magnetic-type fields, temporal updates for the electric fields on the lattice can be deducted and backinterpolated to the charges for the force calculation.

This algorithm is not used as widely as many other well known electrostatics algorithms, such as the various particle-mesh approaches [12]. But it comes with some benefits that have become very interesting over the past few years. Since the Maxwell Equations for electrodynamics are intrinsically local and require no global information on the system, one gains two main advantages:

- First, unlike for all Ewald-based algorithms, the parallelization for such a local system of equations is trivial. And the scaling of the algorithm is only dependent on the lattice mesh size and does therefore scale linearly $\mathcal{O}(N)$ for a fixed particle density. This is a very intriguing feature in a time where massively parallel grid computing and systems in the order of 10^9 charges need to be considered.
- Second, a periodic box geometry can be dealt with very naturally by mathematically connecting each boundary to its oppositely placed boundary and propagating the fields accordingly. Another welcomed feature is that because of its locality the method allows for arbitrary changes of the dielectric properties within the system.

Performance and Precision

As expected from an intrinsically local algorithm, MEMD scales linearly with the mesh size, even to high particle numbers. For homogeneously distributed charges in a dense system (e.g. a molten salt), it outperforms P^{3}M for a comparable precision at about 5,000 and more charges. However, the MEMD algorithm cannot be tuned

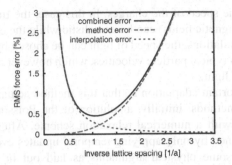

Fig. 2 The two error sources of the algorithm. This graph shows two things: for once, the error cannot be tuned to arbitrarily small values, but only to about 1 %. One can also observe that the method error increases for finer lattice spacings and therefore a smaller field propagation speed. On the other hand, the interpolation error increases if the mesh becomes too coarse

to a given precision, although some statements can be made on the systematic errors. The two main systematic errors stem from the speed of light parameter not being infinite (which would mean perfect electrostatic limit) and from the charge interpolation on the lattice.

The first error is of algorithmic origin and is merely the difference between electrodynamics and the static limit. As can be seen e.g. in Ref. [21], this error scales with $1/c^2$, where c is the speed of light. The second error is larger in comparison and diverges for very fine meshes, since only next neighbor interpolation is performed. It scales with $1/a^3$, where a is the lattice spacing. Since the speed of light (i.e. the propagation speed of the magnetic fields) also directly depends on the lattice spacing, these two errors can be combined to find the best suited mesh for the system (see Fig. 2).

In conclusion one can see that for dense systems or at high particle numbers, the MEMD algorithm provides a very flexible and fast alternative method to calculate Coulomb interactions.

4.3   The ICC⋆ Algorithm for Dielectric Interfaces

Taking into account dielectric interfaces of arbitrary shape in coarse grained simulations is a challenging task. The ICC⋆ algorithm [23, 44] allows to do so with acceptable extra cost and inherent consistency with the desired periodic boundary conditions. At dielectric interfaces the normal component of the electric field has a discontinuity of the following form:

$$\varepsilon_1 E_1 \cdot n = \varepsilon_2 E_2 \cdot n, \qquad (1)$$

where n is the normal vector pointing from regions 2 to 1. This discontinuity can also be interpreted as an effective charge distribution on the surface: the induced charge. For the induced charge density the following equation has to hold:

$$\sigma_{ind} = \frac{\varepsilon_1}{2\pi} \frac{\varepsilon_1 - \varepsilon_2}{\varepsilon_1 + \varepsilon_2} E \cdot n. \tag{2}$$

The idea of the ICC⋆ algorithm is to self-consistently determine this charge distribution in every MD step. The points of a discretized boundary play the role of usual particles, except they are fixed in space. A simple relaxation scheme is applied, where a new guess of the charge density on every boundary element is calculated from

$$\sigma^i_{ind} = (1 - \lambda)\,\sigma^{i-1}_{ind} + \lambda \left(\frac{\varepsilon_1}{2\pi} \frac{\varepsilon_1 - \varepsilon_2}{\varepsilon_1 + \varepsilon_2} E \cdot n \right), \tag{3}$$

where λ is a relaxation parameter, that is typically chosen in the range between 0.5 and 0.9 and E denotes the electric field caused by all other charges. This iteration usually takes only a few steps, because the position of the charged particles in the system changes only slightly.

In many cases just one update per MD steps yields sufficient accuracy. In this case, the extra computational cost of taking into account the dielectric boundary forces is only due to the increased number of charges in the system stemming from the surface discretization. In a system with an intrinsic surface charge, i.e. due to dissociation of surface groups, this process is computationally for free, because the electrostatic field created by the surface charges needs to be calculated anyways.

To create dielectric interfaces ESPResSo offers a set of commands that create an almost equidistant discretization of geometric primitives. This allows to create objects of a very complex geometry. We also plan to combine the algorithm with the concept of rigid bodies as described in Sect. 5. This would allow also to study moving dielectric objects, which is of interest e.g. in colloidal electrophoresis.

5 Rigid Bodies

We have added the ability to simulate rigid bodies built up from particles to ESPResSo. This is useful to model clusters of particles that move as a whole, to build extended bodies which interact with a fluid (like the raspberry model for colloidal particles [27]), and to construct "non-sliding" bonds, which, for instance, attach polymer chains to specific spots on the surface of a colloidal particle. Some of these cases have traditionally been handled by constraint solvers or by very stiff bond potentials. However, our rigid bodies have a smaller computational overhead, and allow for larger aggregates.

The simulation of rigid bodies is implemented in ESPResSo using the concept of virtual sites. These are usual MD particles with the one exception that their position, orientation, and velocity is not obtained from the integration of Newton's equations of motion. Instead, they are calculated from the position and orientation of another particle in the simulation. This non-virtual particle should be located in the center of

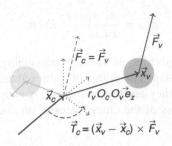

Fig. 3 Illustration of the rigid body implementation. The position x_v of the virtual particle is calculated from the position of the center particle, x_c, its current orientation, O_c, and the virtual particle's relative position in the body frame, $r_v O_v e_z$. A force F_v acting on the virtual site induces the force F_c and the torque T_c

mass of the rigid body and should carry the mass and inertial tensor of the complete body. The position and orientation of this center particle are obtained from the integration of Newton's equation, just as any other normal MD particle. The virtual sites, that give the shape and interactions of the rigid body, are placed relative to the position and orientation of the center particle, and forces acting on them are translated back to forces and torques acting on the center particle. Orientation in this case here refers not just to a director, but to a full local three dimensional reference frame, in which the virtual sites can be placed.

The virtual sites making up the rigid body are placed according to the following rules (compare Fig. 3). The position of a virtual site x_v is obtained from

$$x_v = x_c + r_v O_c O_v e_z, \tag{4}$$

where x_c is the position of the center particle. O_c is the rotation operator that relates the particle-fixed, co-rotating coordinate system of the center particle to the lab frame and represents the orientation of the rigid body. x_c and O_c are the two quantities that are integrated using Newton's equations of motion. The relative position of the virtual site, is represented by O_v, which is the rotation operator which rotates the e_z unit vector such that it becomes co-aligned with the vector connecting the center particle with the virtual site, and by r_v, which is the distance between the virtual particle and the center of mass. In the ESPRESSO package, the rotations O_c and O_v are represented as quaternions.

Accordingly, the velocity of the virtual site is given by

$$v_v = \omega_c \times (x_v - x_c), \tag{5}$$

where ω_c is the rotational velocity of the center particle.

The force on the center of mass due to the force F_v acting on a virtual site is given by

$$F_c = F_v, \tag{6}$$

so that the total force on the center particle j is

$$F_j^{\text{total}} = \sum_{i \text{ virtual site of } j} F_i. \tag{7}$$

The torque generated on the center of mass by a force F_v acting on a virtual site is

$$T_c = (x_v - x_c) \times F_v, \tag{8}$$

so that the total torque acting on the center particle j is

$$T_j^{\text{total}} = \sum_{i \text{ virtual site of } j} (x_i - x_j) \times F_i. \tag{9}$$

In other words, the force acting on the virtual site is copied to the center particle. In addition, the component of the force which is orthogonal to the vector connecting center of mass and virtual site creates a torque on the center of mass.

Combining these formulas, the force calculation proceeds as follows:

1. Place all the particles of the rigid body as virtual sites according to the position and orientation of the center of mass particle (Eqs. (4) and (5)).
2. Calculate the forces in the system. This includes forces between parts of the rigid body (i.e., the virtual sites) and the rest of the system.
3. Collect the forces which have accumulated on the virtual sites and convert them to a force and a torque acting on the center of mass particle (Eqs. (6) and (8)).

Using these forces and torques, the Newton's equations can be integrated for the center of mass of the rigid bodies along with those for all the other non-virtual particles in the system.

6 Dynamic Bonding

For the study of reaction kinetics and irreversible agglomeration, it is necessary that bonds can be created dynamically during the simulation, for example, when particles come closer than a given bonding distance. For this purpose, a collision detection feature has been introduced in ESPRESSO. Breakable bonds, that would be required for studying reaction kinetics of weaker bonds, are at present not implemented, but planned for a future release. Using dynamic bonds is computationally cheaper than using e.g. reactive potentials [8, 17], since one does not need many body potentials, but it is also less natural in terms of atomistic dynamics. For example, our present method does not allow to differentiate between single or double bonds. However, the main aim of ESPRESSO are coarse-grained simulations, where large macromolecules such as soot particles agglomerate. On the size of these particles, the attractive interactions are very short ranged, but strong, which is modeled well by such a discrete on-off bonding, while it is unclear how

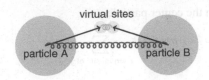

Fig. 4 Construction of the non-sliding bond. Virtual particles are constructed at the point of collision and connected by a zero length bond. The two colliding particles are also directly bound at collision distance

to parametrize e.g. a Tersoff-like potential for macromolecules. Also, our dynamic bonding optionally allows to fix the internal frames of particles against each other, so that the positions of the contacts remains fixed. Again, this makes little sense in terms of atomistic simulations, but is realistic for soot particles, where the sticking is due to permanent deformation of the particles.

When dynamic bonding is turned on, particles that during the simulation come closer than a specified distance get connected by bonds, which at present cannot break again. Therefore, larger and larger clusters are formed. In the simplest case, particles are bonded by a distance-dependent potential like a harmonic or FENE bond. In this case, the particles in the cluster can slide around each other. If the particles do not have a strong repulsive interaction, this might lead to the formation of further bonds within the cluster, resulting in a rather bulky agglomerate. A snapshot of such a cluster can be seen in the left part of Fig. 5. The cluster clearly has a branched structure, which we found to be stable for half a million MD time steps, but the branches are at least three particle diameters wide. For the snapshots, we performed MD simulations of thousand Weeks-Chandler-Andersen (purely repulsive Lennard-Jones) beads in the NVT ensemble, for a total of a million MD steps.

This sliding of connected particles, however, is not desirable in many applications. When, for instance, larger particles collide and stick together due to local surface effects, the particles should always be connected at the particular point where they initially touched. This is achieved by using the rigid body implementation described in Sect. 5: when two particles collide, two virtual sites are created at the point of collision. Each of them is rigidly connected to one of the colliding particles, while these two virtual sites are bound together by a bond potential with an equilibrium distance equal to zero. In addition, a normal bond is formed between the two colliding particles directly (compare Fig. 4). While the bond between the virtual sites prevents the particles from sliding around each other, the bond between the centers of the colliding particle prevents significant motion around the point of collision. The resulting agglomerates (shown in Fig. 5 on the right) have a much more fine and branched structure, with connections as thin as single particles. The depicted picture was stable for also more than 500,000 MD time steps, and initial conditions were the same as for the conventional bonding strategy. Therefore, the finer structure is clearly only due to the non-sliding bonds.

While the current implementation works only on a single CPU, the method is in principle scalable, because at no point information about the complete

Fig. 5 Snapshots of agglomerates generated by dynamic bonding with harmonic bonds (*left*) and with non-sliding bonds (*right*). The structures formed by the non-sliding bonding are finer and more branched, visible by the high amount of voids

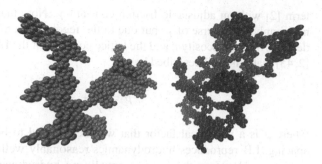

agglomerating cluster is necessary on a single processor, and with time, agglomeration becomes a less and less likely event. We plan to extend the implementation for parallel computing in the near future.

7 Lattice Boltzmann

After 20 years of development the Lattice Boltzmann Method (LBM) has proven to be a powerful tool to investigate dynamics in soft matter systems. In coarse grained simulations it allows to replace the solvent by a mesoscopic fluid, whose dynamics are computed on a discrete lattice. On large length and timescales the LBM leads to a hydrodynamic flow field that satisfies the Navier-Stokes equation. The relaxation of fluid degrees of freedom in liquid systems is much faster than the transport of particles. This separation of timescales allows to neglect the microscopic details of the fluid and is the very reason why different mesoscopic solvent models, such as DPD [19], SRD [29] and the LBM produce equivalent results.

Due to restricted amount of space we only state the properties of the implementation in ESPRESSO and suggest to read the review by Dünweg and Ladd [13] as well as other literature [7, 25, 41, 46]. In ESPRESSO we have implemented a three-dimensional LB model, where velocity space is discretized in 19 velocities (D3Q19). This is sufficient for isothermal flow, as present in almost all soft matter applications. The collision step is performed after a transformation of the populations into mode space as this is necessary for a consistent thermalization [1, 38]. This means, that ESPResSo implements the multiple relaxation time (MRT) LB concept [25]. It also allows to use different relaxation times for different modes, thus to adjust the bulk and shear viscosity of the fluid separately.

As long time scales of fluid relaxation and particle transport are well separated, it is possible to execute the update of the LB fluid less often than the MD steps. Typically one lattice update per five MD steps is sufficient, however, this may depend on other parameters of the system. Besides the update rate, the user conveniently specifies the grid resolution, the dynamic viscosity as well as the density in the usual MD unit system. Additionally an external force density can be applied on the fluid, corresponding to pressure gradients or the effect of gravity. Particles are treated as point particles and are coupled to the LB fluid via a friction term [2] with an adjustable friction constant γ. Note that the particle mobility is not simply the inverse of γ, but due to the feedback from the moving fluid, it also depends on the viscosity η and the lattice constant of the LB lattice. It was shown in [2,45] that it is well described by

$$\mu = \frac{1}{\gamma} + \frac{1}{g\eta a}. \tag{10}$$

where g is a numerical factor that was determined to be ≈ 25 and a is the grid spacing. LB reproduces hydrodynamics reasonably well independent of the grid spacing a. However, there are essentially no hydrodynamic interactions between two particles interacting with the same LB cell, therefore one typically chooses a of the order of the size of the particles.

Boundaries with zero velocity are incorporated easily into the LBM with the *link bounce back* rule introduced in [47]. Boundaries can also have a nonzero velocity. This is achieved by adding an appropriate bias to the reflected population [41]. Compared to other implementations of the LBM, this boundary method is less sophisticated and there are methods that produce hydrodynamic boundaries with higher accuracy and larger robustness. However as ESPRESSO users usually focus on particle dynamics, this is acceptable. The LBM interface offers simple geometric shapes (spheres, cylinders, rhomboids, ...) that can be combined to surfaces of arbitrary complexity. It is fully consistent with the *constraint* concept, that creates extended objects that act as obstacles or also attractive surfaces for MD particles, thus the geometry can easily be made consistent for particles and the fluid. Also the creation of charged or dielectric boundaries (see Sect. 4.3) is consistent.

The computational effort for the LBM scales linearly with the system size. This allows sizes far beyond the possibilities of traditional methods for hydrodynamic interactions in MD simulations, such as Brownian or Stokesian dynamics with the Oseen- or Rotne-Prager-based mobility matrix. The computational effort is considerable, but the local and lattice-based character of the method allows to optimally exploit the computational possibilities of parallel computing and general-purpose graphics processing unit (GPGPU) programming. As very different programming concepts for massive parallel computers and graphics processing units are necessary, the core implementation is separate for both architectures: The implementation for conventional parallel computer applies the MPI communication interface like the rest of ESPRESSO, while the GPU implementation uses the CUDA programming model and language.

As most available LBM implementations use MPI, we only give a brief description of the GPU implementation here, which uses NVIDIA's CUDA framework [33]. GPU computing is based on the idea of executing the same code massively parallel but with different data, i.e. the Single Instruction Multiple Data (SIMD) paradigm. Both steps of the LBM are optimally suited for this scheme, as identical instructions without conditionals are executed on every node. These instructions are performed in many parallel threads on the Streaming Multiprocessors (SM) of the GPU. The LB fluid resides in the video ram of the GPU. The MD code itself is not altered and still running on the CPU. Only in appropriate intervals the particle positions and

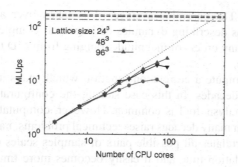

Fig. 6 Scaling of the LB speed with the number of CPU cores (*solid lines*), compared to a single NVIDIA Tesla C2050 GPU for three different lattice sizes (*dashed lines*). The speed is measured in million lattice node updates per second (MLUps). The *dotted line* marks the ideal linear scaling of the speed with the number of cores

velocities are transferred to the GPU to calculate the interaction with the LB fluid, and the resulting forces are transferred back. Due to the dynamic execution order of the threads, schemes like the double buffering method or using atomic operations are essential to avoid race conditions between concurrently running threads.

A comparison of the computing time of the fluctuating LB fluid on a single NVIDIA Tesla C2050 GPU and an AMD CPU cluster with 1.9 GHz Opteron processors is shown in Fig. 6. The size of the GPU memory of 3 GB limits the maximum lattice size to 240^3. For large lattices and few processors the CPU code scales nearly ideally with the system size, while small lattices and many processors result in a large communication overhead, and therefore longer computational times. The speedup of using a state-of-the-art GPU is however striking: For all lattices that are small enough to fit into the GPU memory, the performance of a single NVIDIA GPU can not be reached with the AMD Opteron CPU cluster, no matter how many CPUs are used. The reason is the communication overhead, which becomes dominating at a performance that is a factor of 5–20 below the single GPU, while using more than 50 cores of recent processors.

8 Correlator

Time correlation functions are ubiquitous in statistical mechanics and molecular simulations when dynamical properties of many-body systems are concerned. The velocity autocorrelation function, $\langle \mathbf{v}(t) \cdot \mathbf{v}(t + \tau) \rangle$ which is used in the Green-Kubo relations is a typical example. In general, time correlation functions are of the form

$$C(\tau) = \langle A(t) \otimes B(t + \tau) \rangle, \tag{11}$$

where t is time, τ is the lag time between the measurements of the (vector) observables A and B, and $\otimes$ is an operator which produces the vector quantity

C from A and B. The ensemble average $\langle \cdot \rangle$ is taken over all time origins t. Correlation functions describing dynamics of large and complex molecules such as polymers span many orders of magnitude, ranging from MD time step up to the total simulation time.

It is trivial to compute a correlation function which spans a short time scale, say less than three decades. In this case storing the configurations in a file and using an external analysis tool is common. However, computation of correlation functions which span many decades raises technical problems, namely: (1) A trivial algorithm which correlates all possible pairs of samples scales quadratically with the maximum correlation time and it easily becomes more time-consuming than the actual simulation; (2) Storing configurations too often (each few time steps) produces significant read/write overhead both at the simulation time and post-processing and requires enormous storage space; (3) Specifically for ESPRESSO, storing configurations can only be done at the script level and produces additional overhead when performed too often. Problem 1 can be resolved by using an efficient correlation algorithm; problems 2 and 3 can be resolved by correlating on the fly, without storing configurations too often and passing control to the scripting interface and back. An apparent drawback is that the post-processing of data is no longer possible and the simulation has to be re-done if a new correlation should be computed. However, repeating the simulation is often less computationally expensive than reading the positions of the particles from disk again, due to the large amounts of data.

Since version 3.1 ESPRESSO features an interface for efficient computation of time correlation functions. First, the user has to define an observable at the script level, which creates the corresponding variable in the kernel and makes it available to the correlator. In the next step, he defines which observables shall be correlated, what is the correlation operation, sampling rate (minimum lag time) and the maximum lag time. Optionally, the correlation estimates (and the respective observables) can be updated automatically without the need for an explicit update call at the script level. Furthermore, parameters of the correlation algorithm can be chosen as described below, which influence the effort needed to compute the correlations as well as the statistical quality of the result. In addition, the correlator can also process data input from the scripting interface or from a file, which enables ESPRESSO to be used as an efficient correlator for data produced by other programs.

Algorithm: Multiple Tau Correlator

Here we briefly sketch the multiple tau correlator which is implemented in ESPRESSO. For a more detailed description and discussion of the properties of the algorithm concerning parameter values, statistical and systematic errors, we refer the reader to a recent excellent paper by Ramírez et al. [36]. This type of correlator has been used for years in dynamic light scattering [37]. As a special case, its application to the velocity autocorrelation function is described in detail in the textbook of Frenkel and Smit [15]. Nevertheless, despite its obvious advantages, it has been used scarcely by the simulation community.

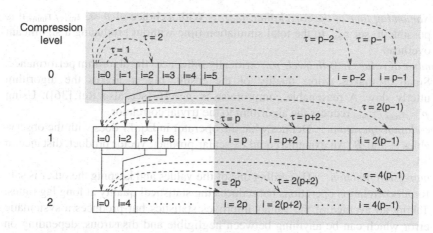

Fig. 7 Schematic representation of the correlator algorithm. Values of i indicate the time of the measurement. *Dashed lines* show values which are correlated and the corresponding lag times. *Solid lines* show values which are merged together in one compression step. At the lowest level, all possible pairs of entries are correlated. At higher levels, only entries separated by more than $p/2$ values are correlated. This is indicated by the *gray shaded* background

The algorithm of the multiple tau correlator is schematically shown in Fig. 7. Index i in the figure denotes an observable which was measured at time $i\,\tau_{min}$ where τ_{min} is the sampling interval. We further drop τ_{min} to simplify the notation. The main idea of the multiple tau correlator is to correlate all possible pairs of observables within the relatively short interval of lag times, $[0 : (p-1)]$, which we refer to as compression level 0. For lag times greater than $p-1$, we first apply a compression to the original data in order to produce compression level 1. In the compression step, one value which goes to the higher level is produced from two values in the lower level, e.g. by taking an average of the two or by discarding one of them. In compression level 1, data with lag times $[p : 2(p-1)]$ are correlated. Data with lag times $[2p : 4(p-1)]$ are correlated in compression level 2 and so on. The number of correlated values at each level is constant but the lag time and the time span of the level increases by a factor of 2. Hence the computational effort increases only logarithmically with the maximum lag time. Thus adding one decade in the lag times increases the effort by a constant amount. The same holds for the amount of memory required to store the compressed history. In the implementation, each compression level is a buffer with $(p + 1)$ entries. When a new value arrives to the full buffer, two values are pushed to the higher level.

There are several relevant parameters which influence the performance of the correlation algorithm. Their influence on statistical quality of the data has been critically examined in the paper by Ramírez et al. [36]. Here we just state the main points. The following parameters are most important:

Sampling interval, τ_{min}. When set to MD time step, it produces a noticeable overhead in computational time. When set to more than ten MD steps, the overhead becomes negligible.

Maximum lag time of the correlation, τ_{max}.   With a choice of $p \ll \tau_{max}/\tau_{min}$, it is possible to set τ_{max} to the total simulation time without producing a significant overhead.

Number of values in 0-th level, p,   critically influences the algorithm performance. Small p makes worse quality of the result, big p makes the algorithm utterly slow. A reasonable compromise is $p = 16$ (see also Ref. [36]). Using $p = \tau_{max}/\tau_{min}$, reduces the algorithm to the trivial correlator.

Correlation operation.   Defines what the operator in Eq. 11 does with the observables A and B. An example could be scalar product, vector product, distance in 3d, . . .

Compression function.   Discarding one of the values and keeping the other is safe for all correlation operations but reduces the statistical quality at long lag times. Taking an average of the values improves statistics but produces a systematic error which can be anything between negligible and disastrous, depending on the physical properties of the correlation function and other parameters of the algorithm.

With this algorithm we have been able to compute on the fly correlations spanning eight decades in the lag times. The overhead in computer time produced by the correlations was up to a factor of 2 when many correlations were computed simultaneously with τ_{min} equal to MD time step and $p = 16$. The overhead was caused mainly by frequent updates of the observables based on system configuration and could be largely reduced by taking a longer sampling interval.

9   Conclusions and Outlook

To summarize our contribution, we have described some recent additions to our ESPRESSO software package, Version 3.1. Compared to the last version that was published in [26], a number of new features have been added which make ESPRESSO a unique software for coarse-grained simulations. It can now treat systems with varying dielectric media, and has the first implementation of the scalable MEMD algorithm for electrostatic interactions. We have included rigid bodies and dynamic bond formation for studying agglomeration. Also, the hydrodynamic interactions of a coarse-grained particle within an implicit solvent can be handled via the Lattice Boltzmann method.

Our LB solver can use the computing power of GPUs, of which a single card is in most cases sufficient to replace a compute cluster. We plan to make this impressive speedup available for other time-consuming parts of the simulation, namely the electrostatic algorithms P^3M and ELC. Also, GPUs finally provide enough computational power to allow for solving the Poisson-Boltzmann equations on the fly, which makes it possible to study electrokinetic effects even in complex multi-particle systems on realistic time scales.

Unfortunately, the Tcl scripting language that provides the user interface for ESPRESSO has come to age and is slow when it comes to numerical computation. Although most of the ESPRESSO core is written in C, this affects many users, who implement their own analysis routines or other computations in Tcl. Therefore, we have decided to switch ESPRESSO's user interface from Tcl to Python in one of the upcoming versions of the software. The more modern scripting language Python[12] has drawn a lot of attention to itself over the last decade and is widely embraced by the scientific community.[13] It provides many useful packages, for example for scientific computing, for two- and three dimensional plotting, for statistical analysis, for visualization, etc., which will then be easily available to the ESPRESSO users.

In addition, we and several other groups contributing to ESPRESSO are planning many more improvements in the future, and we hope to attract more users to actively contribute to the code. The current status of the package and the latest code can be found on our web site http://espressomd.org, which serves as our collaborative development platform.

Acknowledgements We want to thank the more than 20 researchers, that have contributed to the ESPRESSO code so far. Without these contributions and, of course, the feedback of our users, ESPRESSO would never have reached its current status.

References

1. R. Adhikari, K. Stratford, M.E. Cates, A.J. Wagner, Fluctuating Lattice Boltzmann. Europhys. Lett. **71**, 473 (2005)
2. P. Ahlrichs, B. Dünweg, Simulation of a single polymer chain in solution by combining lattice boltzmann and molecular dynamics. J. Chem. Phys. **111**, 8225–8239 (1999)
3. A. Arnold, C. Holm, MMM2D: a fast and accurate summation method for electrostatic interactions in 2d slab geometries. Comput. Phys. Commun. **148**, 327–348 (2002)
4. A. Arnold, C. Holm, Efficient methods to compute long range interactions for soft matter systems, in *Advanced Computer Simulation Approaches for Soft Matter Sciences II*, ed. by C. Holm, K. Kremer. Advances in Polymer Sciences, vol. II (Springer, Berlin, 2005), pp. 59–109
5. A. Arnold, J. de Joannis, C. Holm, Electrostatics in periodic slab geometries I. J. Chem. Phys. **117**, 2496–2502 (2002)
6. A. Arnold, B.A. Mann, H. Limbach, C. Holm, *ESPResSo – An Extensible Simulation Package for Research on Soft Matter Systems*, ed. by K. Kremer, V. Macho. Forschung und wissenschaftliches Rechnen 2003, GWDG-Bericht, vol. 63 (Gesellschaft für wissenschaftliche Datenverarbeitung mbh, Göttingen, 2004), pp. 43–59
7. R. Benzi, S. Succi, M. Vergassola, The lattice Boltzmann equation: Theory and applications. Phys. Rep. **222**, 145–197 (1992)
8. D.W. Brenner, O.A. Shenderova, J.A. Harrison, S.J. Stuart, B. Ni, S.B. Sinnott, A second-generation reactive empirical bond order (REBO) potential energy expression for hydrocarbons. J. Phys. Condens. Matter **14**, 783 (2002)

---

[12]http://python.org

[13]http://www.scipy.org/

9. J.J. Cerdà, V. Ballenegger, O. Lenz, C. Holm, P3M algorithm for dipolar interactions. J. Chem. Phys. **129**, 234104 (2008)
10. M. Deserno, C. Holm, How to mesh up Ewald sums I: a theoretical and numerical comparison of various particle mesh routines. J. Chem. Phys. **109**, 7678 (1998)
11. M. Deserno, C. Holm, How to mesh up Ewald sums II: an accurate error estimate for the Particle-Particle-Particle-Mesh algorithm. J. Chem. Phys. **109**, 7694 (1998)
12. M. Deserno, C. Holm, S. May, The fraction of condensed counterions around a charged rod: comparison of Poisson-Boltzmann theory and computer simulations. Macromolecules **33**, 199–206 (2000)
13. B. Dünweg, A.J.C. Ladd, Lattice Boltzmann simulations of soft matter systems, in *Advanced Computer Simulation Approaches for Soft Matter Sciences III*, ed. by C. Holm, K. Kremer. Advances in Polymer Science, vol. 221 (Springer, Berlin, 2009), pp. 89–166
14. P. Español, P. Warren, Statistical mechanics of dissipative particle dynamics. Europhys. Lett. **30**, 191 (1995)
15. D. Frenkel, B. Smit, *Understanding Molecular Simulation*, 2nd edn. (Academic, San Diego, 2002)
16. J.G. Gay, B.J. Berne, Modification of the overlap potential to mimic a linear site-site potential. J. Chem. Phys. **74**, 3316–3319 (1981)
17. M. Griebel, J. Hamaekers, F. Heber, A molecular dynamics study on the impact of defects and functionalization on the young modulus of boron-nitride nanotubes. Comput. Mater. Sci. **45**, 1097–1103 (2009)
18. B. Hess, C. Kutzner, D. van der Spoel, E. Lindahl, GROMACS 4: algorithms for highly efficient, load-balanced, and scalable molecular simulation. J. Chem. Theory Comput. **4**, 435–447 (2008)
19. P.J. Hoogerbrugge, J.M.V.A. Koelman, Simulating microscopic hydrodynamic phenomena with dissipative particle dynamics. Europhys. Lett. **19**, 155–160 (1992)
20. W. Humphrey, A. Dalke, K. Schulten, VMD: visual molecular dynamics. J. Mol. Graph. **14**, 33–38 (1996)
21. J.D. Jackson, *Classical Electrodynamics*, 3rd edn. (Wiley, New York, 1999)
22. C. Junghans, S. Poblete, A reference implementation of the adaptive resolution scheme in ESPResSo. Comput. Phys. Commun. **181**, 1449–1454 (2010)
23. S. Kesselheim, M. Sega, C. Holm, Applying ICC to DNA translocation: effect of dielectric boundaries. Comput. Phys. Commun. **182**, 33–35 (2011)
24. A. Kolb, B. Dünweg, Optimized constant pressure stochastic dynamics. J. Chem. Phys. **111**, 4453–4459 (1999)
25. P. Lallemand, D. d'Humieres, L.S. Luo, R. Rubinstein, Theory of the lattice Boltzmann method: three-dimensional model for linear viscoelastic fluids. Phys. Rev. E **67**, 021203 (2003)
26. H.J. Limbach, A. Arnold, B.A. Mann, C. Holm, ESPResSo – an extensible simulation package for research on soft matter systems. Comput. Phys. Commun. **174**, 704–727 (2006)
27. V. Lobaskin, B. Dünweg, A new model for simulating colloidal dynamics. New J. Phys. **6**, 54 (2004)
28. A.C. Maggs, V. Rosseto, Local simulation algorithms for Coulombic interactions. Phys. Rev. Lett. **88**, 196402 (2002)
29. A. Malevanets, R. Kapral, Continuous-velocity lattice-gas model for fluid flow. Europhys. Lett. **44**, 552 (1998)
30. N.S. Martys, R.D. Mountain, Velocity verlet algorithm for dissipative-particle-dynamics-based models of suspensions. Phys. Rev. E **59**, 3733–3736 (1999)
31. R. Messina, Effect of image forces on polyelectrolyte adsorption at a charged surface. Phys. Rev. E **70**, 051802 (2004)
32. MPI Forum, MPI: *A Message-Passing Interface* (MPI) *Standard Version 1.3* (2008)
33. NVIDIA Corporation, *NVIDIA CUDA reference manual version 3.2* (2010)
34. I. Pasichnyk, B. Dünweg, Coulomb interactions via local dynamics: a molecular-dynamics algorithm. J. Phys. Condens. Matter **16**, 3999–4020 (2004)

35. J.C. Phillips, R. Braun, W. Wang, J. Gumbart, E. Tajkhorshid, E. Villa, C. Chipot, R.D. Skeel, L. Kalé, K. Schulten, Scalable molecular dynamics with NAMD. J. Comput. Chem. **26**, 1781–1802 (2005)
36. J. Ramirez, S.K. Sukumaran, B. Vorselaars, A.E. Likhtman, Efficient on the fly calculation of time correlation functions in computer simulations. J. Chem. Phys. **133**, 154103 (2010)
37. K. Schätzel, M. Drewel, S. Stimac, Photon-correlation measurements at large lag times – improving statistical accuracy. J. Mod. Opt. **35**, 711–718 (1988)
38. U.D. Schiller, Thermal fluctuations and boundary conditions in the lattice Boltzmann method. Ph.D. thesis, Johannes Gutenberg-Universität Mainz, Fachbereich 08: Physik, Mathematik und Informatik (2008)
39. E.R. Smith, Electrostatic potentials for thin layers. Mol. Phys. **65**, 1089–1104 (1988)
40. T. Soddemann, B. Dünweg, K. Kremer, Dissipative particle dynamics: a useful thermostat for equilibrium and nonequilibrium molecular dynamics simulations. Phys. Rev. E **68**, 46702 (2003)
41. S. Succi, *The Lattice Boltzmann Equation for Fluid Dynamics and Beyond* (Oxford University Press, New York, 2001)
42. S. Tyagi, A. Arnold, C. Holm, ICMMM2D: an accurate method to include planar dielectric interfaces via image charge summation. J. Chem. Phys. **127**, 154723 (2007)
43. S. Tyagi, A. Arnold, C. Holm, Electrostatic layer correction with image charges: a linear scaling method to treat slab 2d + h systems with dielectric interfaces. J. Chem. Phys. **129**, 204102 (2008)
44. C. Tyagi, M. Süzen, M. Sega, M. Barbosa, S. Kantorovich, C. Holm, An iterative, fast, linear-scaling method for computing induced charges on arbitrary dielectric boundaries. J. Chem. Phys. **132**, 154112 (2010)
45. O.B. Usta, A.J.C. Ladd, J.E. Butler, Lattice-Boltzmann simulations of the dynamics of polymer solutions in periodic and confined geometries. J. Chem. Phys. **122**, 094902 (2005)
46. D.A. Wolf-Gladrow, *Lattice-Gas Cellular Automata and Lattice Boltzmann Models: An Introduction*, vol. 1725 (Springer, New York, 2000)
47. D.P. Ziegler, Boundary conditions for lattice Boltzmann simulations. J. Stat. Phys. **71**, 1171–1177 (1993)

On the Rate of Convergence of the Hamiltonian Particle-Mesh Method

Bob Peeters, Marcel Oliver, Onno Bokhove, and Vladimir Molchanov

Abstract The Hamiltonian Particle-Mesh (HPM) method is a particle-in-cell method for compressible fluid flow with Hamiltonian structure. We present a numerical short-time study of the rate of convergence of HPM in terms of its three main governing parameters. We find that the rate of convergence is much better than the best available theoretical estimates. Our results indicate that HPM performs best when the number of particles is on the order of the number of grid cells, the HPM global smoothing kernel has fast decay in Fourier space, and the HPM local interpolation kernel is a cubic spline.

Keywords Hamiltonian particle-mesh method • Rate of convergence • Numerical tests

1 Introduction

The Hamiltonian Particle-Mesh (HPM) method is a particle-in-cell method for compressible fluid flow with the special property that the discrete equations of motion form a Hamiltonian N-particle mechanical system. It was originally proposed in the context of the shallow water equations by Frank et al. [6], and tested on a variety of two-dimensional geophysical flow problems [3–5]. Moreover, the HPM method was shown to be convergent [10, 11] as the number of particles N tends to infinity.

Comparing the HPM method with classical smoothed particle hydrodynamics (SPH) which also possesses a Hamiltonian structure and associated conservation

B. Peeters · O. Bokhove
Department of Applied Mathematics, University of Twente, Enschede, The Netherlands

M. Oliver (✉) · V. Molchanov
School of Engineering and Science, Jacobs University, 28759 Bremen, Germany
e-mail: oliver@member.ams.org

M. Griebel and M.A. Schweitzer (eds.), *Meshfree Methods for Partial Differential Equations VI*, Lecture Notes in Computational Science and Engineering 89, DOI 10.1007/978-3-642-32979-1_2, © Springer-Verlag Berlin Heidelberg 2013

laws [7, 9, 13], we note that the known convergence results are very similar in the sense that both can be shown to be of order $O(N^{2-\varepsilon})$ for any $\varepsilon > 0$ provided the underlying kernel functions satisfy certain technical conditions [11]. On the other hand, one time step of the HPM method can be computed in $O(N)$ or $O(N \ln N)$ operations. The computational complexity of SPH is algebraically superlinear in N as the number of interactions per particle must grow without bound as $N \to \infty$. These facts suggest that HPM may be regarded as a "fast" implementation of the SPH algorithm.

In addition to the number of particles N, the HPM method has two length scale parameters—the size λ of the auxiliary mesh, and a smoothing length scale μ. A priori, these three parameters may be chosen independently. The theory in [11] yields a relation for the scaling of λ and μ as a function of N to achieve a certain rate of convergence; however, it is not known if this relation is optimal. Further, it requires technical conditions on the kernel functions not all of which are met in [6]. Even though satisfactory error behavior is frequently reported, this has not yet been investigated systematically. Finally, it is not known whether the grid itself has a positive impact on the scheme beyond facilitating fast computation of long-range interactions. Thus, the purpose of this paper is to take a very simple, well-controlled numerical setting to study the behavior of HPM as a function of its governing parameters and of the relevant properties of the HPM kernels, and to compare its performance to the classical SPH algorithm.

In this paper, we restrict ourselves to studying the short-time behavior of the HPM method. This corresponds to the analytic results in [11], where the error constants may grow exponentially in time as is typical for general trajectory error estimates for evolution equations. Thus, the Hamiltonian aspects of HPM shall not be considered further. Moreover, we restrict ourselves to two known exact solutions as test cases, namely Burgers' solution in space dimension one and the Iacono cosine vortex in space dimension two. The former is a special solution to the plain irrotational shallow water equations; for the latter, we must include the non-inertial effects of rotation as well as nontrivial bottom topography. Burgers' solution is considered only up to before the time of shock formation. It tests the ability of the particle scheme to cope with increasingly inhomogeneous particle distributions. The cosine vortex is an Eulerian steady state with Lagrangian particle velocities of order one. As such, it serves as a prototype of an optimally well-behaved exact solution which yet poses nontrivial challenges to a particle scheme.

We find that the rate of convergence is much better than the best available theoretical estimates in [11]. Further, our results indicate that HPM performs best when the number of particles is on the order of the number of grid cells, the HPM global smoothing kernel has fast decay in Fourier space, and the HPM local interpolation kernel is a cubic spline.

The outline of the paper is as follows. Section 2 introduces the continuum shallow water equations and their discretization with the HPM method. Section 3 states a simplified version of the convergence result of [11]. Section 4 introduces the exact special solutions of the shallow water equation that we benchmark against.

The numerical results are detailed in Sect. 5, and the paper concludes with a brief discussion of these results in Sect. 6.

2 The HPM Method for Shallow Water

We apply the HPM method to the rotating shallow water equations with bottom topography. On the one hand, the rotating shallow water equations can be seen as a simple example of a barotropic flow. On the other hand, the Coriolis force and bottom topography terms allow for nontrivial exact two-dimensional steady states which we will use as one of our benchmarks.

The continuum equations of motions describe the evolution of a d-dimensional velocity field $u = u(x, t)$ and a density field $\rho = \rho(x, t)$ via

$$\partial_t u + u \cdot \nabla u + \mathbb{J}u = -\nabla(\rho + b), \tag{1a}$$

$$\partial_t \rho + \nabla \cdot (\rho u) = 0, \tag{1b}$$

where $\mathbb{J}$ is a zero-order skew-symmetric operator, $b = b(x)$ is a smooth, time-independent function, and $x = (x_1, \ldots, x_d)$. When considered on a two-dimensional spatial domain, then u describes the evolution of the vertically averaged velocity field, ρ describes the layer depth, and b the spatial variation of the bottom topography of a thin layer of fluid. In the shallow water literature, the layer depth is usually denoted h. Here, as in [11], we disregard the physical connotations and write h to denote the numerical approximation to ρ. More generally, the forcing term $\nabla \rho$ arises from the choice of barotropic fluid pressure $p(\rho) = \rho^2/2$.

In our experiments, we take $d = 1, 2$ and supply periodic boundary conditions on $\mathbb{T}^d \equiv [-\pi, \pi)^d$. When $d = 1$, we take $\mathbb{J} = 0$; when $d = 2$, we use the standard so-called f-plane Coriolis term $\mathbb{J}u = (-u_2, u_1)$. The physical constant of gravity and the Coriolis constant have been set to unity.

To define the HPM method, we introduce a regular grid with K nodes in each dimension. The locations of the mesh nodes are given by $\{x_\alpha \equiv \lambda\alpha : \alpha \in \mathbb{G}^d\}$ on $\mathbb{T}^d$, where $\mathbb{G} = \mathbb{Z} \cap [K/2, K/2)$ is the index set, always interpreted in modulo K arithmetic, and $\lambda = 2\pi/K$ is the mesh size.

We first define a *local partition of unity kernel* via a compactly supported shape function Ψ. Here, we restrict ourselves to considering tensor-product B-splines of varying order. We note that the spline of order r satisfies a so-called Strang–Fix condition of order $p = r + 1$ which expresses that polynomials of degree less than p can be composed of integer translates of the shape function Ψ; it plays a crucial role in the analysis in [11]. All results in the following are labeled by the order p of the Strang–Fix condition used. Once the shape function Ψ is specified, the scaled kernel

$$\psi_\lambda(x) = \lambda^{-d} \Psi(x/\lambda) \tag{2}$$

and its translates form a periodic partition of unity on the mesh.

Second, we define a *global smoothing operator* via discrete convolution on the mesh as follows. For a mesh function $h = (h_\alpha)_{\alpha \in \mathbb{G}^d}$, the action of the smoothing operator $S_{\lambda,\mu}$ on h at grid node $\alpha \in \mathbb{G}^d$ is computed by filtering high frequencies in discrete Fourier space via

$$(S_{\lambda,\mu}h)_\alpha = \sum_{\gamma \in \mathbb{G}^d} e^{i\gamma \cdot x_\alpha} \frac{1}{(1 + |\mu\gamma|^2)^q} \tilde{h}(\gamma), \tag{3}$$

where μ defines a *global smoothing length scale* and

$$\tilde{h}(\gamma) = \frac{1}{K^d} \sum_{\beta \in \mathbb{G}^d} e^{-i\gamma \cdot x_\beta} h_\beta \tag{4}$$

denotes the discrete Fourier transform of h. In other words, $S_{\lambda,\mu}$ is a discrete approximation to the q-th power of the inverse Helmholtz operator $1 - \mu^2 \Delta$.

The HPM approximation [6] to (1) is a set of ordinary differential equations describing the motion of N fluid particles of mass m_k at positions X_N^k, where $k = 1, \ldots N$, via

$$\frac{d}{dt} X_N^k(t) = U_N^k(t), \tag{5a}$$

$$\frac{d}{dt} U_N^k(t) = -\mathbb{J}U_N^k(t) - \nabla(\bar{h}(x,t) + \bar{b}(x))\Big|_{x = X_N^k(t)}, \tag{5b}$$

where the smoothed layer depth $\bar{h}$ and smoothed bottom topography $\bar{b}$ are computed from the finite ensemble of particles in a three-step process. First, we obtain an interpolated layer depth on the grid via

$$h_\alpha(t) = \sum_{k=1}^{N} m_k \, \psi_\lambda(x_\alpha - X_N^k(t)). \tag{6a}$$

Second, we introduce a smoothed layer depth on the grid,

$$\bar{h}_\alpha(t) = (S_{\lambda,\mu}h)_\alpha, \tag{6b}$$

where $h = (h_\alpha)_{\alpha \in \mathbb{G}^d}$ denotes the layer depth approximation on the grid. Third, we interpolate the layer depth field from the grid onto the entire domain by using the partition of unity kernels from (2) once more, setting

$$\bar{h}(x,t) = \lambda^d \sum_{\alpha \in \mathbb{G}^d} \bar{h}_\alpha(t) \, \psi_\lambda(x - x_\alpha). \tag{6c}$$

Similarly, the bottom topography contribution is computed via

$$\bar{b}_\alpha = (S_{\lambda,\mu} b)_\alpha \tag{7a}$$

where, in abuse of notation, we use b to also denote the topography function evaluated on the grid, and set

$$\bar{b}(x) = \lambda^d \sum_{\alpha \in \mathbb{G}^d} \bar{b}_\alpha \, \psi_\lambda(x - x_\alpha) . \tag{7b}$$

Expressions (6c) and (7b) can now be analytically differentiated and used on the right hand side of the discrete momentum equation (5b). The reason for using the same interpolation kernel ψ_λ in (6a) and in (6c) is that only then the HPM dynamics has a Hamiltonian structure [6], which can be seen as a direct discrete analog of the parcel Hamiltonian continuum formulation of [2]. This, however, will not play any further role in this paper.

We remark that our scaling of the masses m_k in (6a) is that commonly used in the SPH literature. It differs from the definition of m_k and ψ_λ used in [6]. In our scaling, sums over the grid such as appear in (6c) can be read as a Riemann sum approximation of an integral over the torus where λ^d is the volume of a single cell. Moreover, we can cleanly discuss the SPH limit of HPM, namely the limit that $\lambda \to 0$ with N and μ fixed (the m_k as defined in [6] become singular in this limit).

In our tests, we initially place L particles in each dimension on a regular particle mesh, so that $N = L^d$ and the initial particle positions X_N^k can be identified, by enumeration, with the initialization mesh points $\{X_\beta \equiv \Lambda\beta : \beta \in \mathbb{H}^d\}$ where $\mathbb{H} = 1/2 + \mathbb{Z} \cap [-L/2, L/2)$ with particle mesh spacing $\Lambda = 2\pi/L$.

Then, with the same identification between enumeration index k and multi-index β, we set, at time $t = 0$, m_β to be an approximation to the mass in cell β given by the d-dimensional trapezoidal rule

$$m_\beta = \frac{\Lambda^d}{2^d} \sum_\gamma \rho(X_\beta + \tfrac{\Lambda}{2}\gamma, 0) \approx \int_{X_\beta + [-\Lambda/2, \Lambda/2]^d} \rho(a)\, da , \tag{8}$$

where the sum ranges over all possible d-vectors γ whose entries are 1 or -1.

3 Theoretical Estimates and Consequences

We measure the error in terms of the L^2-like error functional

$$Q = \frac{1}{2} \sum_{k=1}^N m_k \left| U_N^k - u(X_N^k) \right|^2 + \frac{\lambda^d}{2} \sum_{\alpha \in \mathbb{G}^d} \left| (S_{\lambda,\mu}^r h_{\text{tot}})_\alpha - \rho_{\text{tot}}(x_\alpha) \right|^2$$

$$\equiv Q_{\text{kin}} + Q_{\text{pot}} \tag{9}$$

where $h_{tot} = h+b$ and $\rho_{tot} = \rho+b$ are the approximate and exact total layer depths, respectively, and $S^r_{\lambda,\mu}$ denotes the convolution square root of $S_{\lambda,\mu}$ which acts, due to the Fourier convolution theorem, on the grid function h via

$$(S^r_{\lambda,\mu}h)_\alpha = \sum_{\gamma \in \mathbb{G}^d} e^{i\gamma \cdot x_\alpha} \frac{1}{(1+|\mu\gamma|^2)^{q/2}} \tilde{h}(\gamma). \tag{10}$$

The error functional can be seen as a direct generalization of the HPM Hamiltonian [6] where the convolution square root arises from symmetrizing the expression for the potential energy. Our error functional is also motivated by the one used in [12], where the integral over the layer depth error used there is replaced by its grid-approximation here. Under mild smoothness assumptions, the difference between the two error measures vanishes as $\lambda \to 0$.

Vanishing of the error functional Q as $N \to \infty$ corresponds to convergence of the numerical solution of (5)–(7) to a solution of (1). The following theorem states sufficient conditions for convergence and an associated error bound.

Theorem 1 ([11]). *Suppose that the shallow water equations possess a classical solution of class*

$$u, \rho \in C^1([0,T]; H^{4+d}(\mathbb{T}^d)) \tag{11}$$

for some $T > 0$. Apply the Hamiltonian particle-mesh method with a tensor-product cardinal B-spline of order $p - 1 \geq 2$ as local interpolation kernel and with (3) as the global smoothing operator. Then, with $q > p + d/2$,

$$\lambda \sim N^{-\frac{1}{d}}, \quad and \quad \mu \sim L^{-\frac{p-1}{p+2+d}} \tag{12}$$

as $N \to \infty$, there exist constants C_1 and C_2 such that

$$Q(t) \leq C_1 e^{C_2 t} L^{-\frac{2(p-1)}{p+2+d}} \tag{13}$$

for all $t \in [0,T]$.

A few remarks are in order. First, the theorem was proved for a flat bottom and without rotation; there is, however, no principle obstacle to obtaining a corresponding result in the presence of nontrival topography.

Second, the condition $q > p + d/2$ excludes the bi-Helmholtz operator used in most of the previous work on HPM. This condition is not sharp, but is technically convenient as it ensures a strict ordering of the various contributions to the total error in the proof of the theorem. Hence, we expect that as q is decreased, the rate of convergence in (13) decreases as well.

Third, the rate of convergence in (13) can be improved to any exponent less than two by choosing a local interpolation kernel which satisfies a sufficiently high order of polynomial reproduction, and subject to choosing q large enough.

Finally, it is possible to let the mesh size $\lambda \to 0$ while N and μ remain fixed. Although this limit is not computationally relevant as the computational cost tends to infinity while the accuracy remains finite, it is the limit in which the HPM method tends to the classical SPH algorithm while the sum in the potential energy term of the error functional tends to an integral, so that the error functional converges to the one used by Oelschläger in his proof of convergence of the SPH method [12]. The result is therefore of theoretical interest and can be stated as follows.

Theorem 2 ([11]). *In the setting of Theorem 1, suppose that the shallow water equations are solved by the SPH method with periodic kernel given by (3) with $q > \max\{3 + d/2, d\}$. Then there exist constants C_1 and C_2 such that, letting*

$$\mu \sim N^{-\frac{q-d}{q+2}} \tag{14}$$

as $N \to \infty$, we obtain the error bound

$$Q(t) \le C_1\, e^{C_2 t}\, N^{-\frac{2(q-d)}{d(q+2)}} \tag{15}$$

for all $t \in [0, T]$.

In particular, the order of convergence in (15) can be raised to any exponent less than two by choosing q large enough.

4 Special Solutions

4.1 Burgers' Solution for $d = 1$

We consider the one-dimensional, non-rotating shallow water equations without bottom topography,

$$\partial_t u + u\, \partial_x u = -\partial_x \rho \quad \text{and} \quad \partial_t \rho + u\, \partial_x \rho = -\rho\, \partial_x u, \tag{16}$$

for $x \in [-\pi, \pi)$ with periodic boundary conditions. This system has an analytical solution in terms of the function $J(x, t) = K - 3\sqrt{\rho(x, t)}$ satisfying the inviscid Burgers' equation

$$\partial_t J + J\, \partial_x J = 0 \tag{17}$$

provided that the Riemann invariant $u + 2\sqrt{\rho}$ initially equals the constant K. We choose

$$\rho(x, 0) = \tfrac{1}{9}(K + \sin x)^2 \quad \text{and} \quad u(x, 0) = K - 2\sqrt{\rho(x, 0)} \tag{18}$$

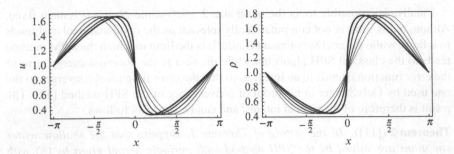

Fig. 1 Burgers' solution of the one-dimensional shallow water equations at times $t = 0, 0.25,$ $0.5, 0.75, 0.95$

with $K = 3$. This solution develops a discontinuity at the earliest time the characteristics of Burgers' equation intersect, so that

$$t_{\text{shock}} = -1/\min \partial_x J(x, 0) = 1 . \tag{19}$$

The steepening of the wave in time is shown in Fig. 1. Since HPM does not contain shock handling, we are not interested in the behavior at or beyond the singularity. Rather, we perform a multi-parameter study of the behavior of HPM on the time interval $0 \leq t \leq 0.95$, which is close enough to the time of shock solution that the final particle distribution is very nonuniform. This example has also been used in [1, 16].

4.2 Cosine Vortex over Topography for $d = 2$

The cosine vortex in our test case belongs to the family of Iacono vortex solutions [8], special Eulerian steady-state solutions to the two-dimensional shallow water equations. These steady-states are interesting as they pose a nontrivial challenge to Lagrangian methods.

Following [8], we impose that the flow is steady with divergence-free horizontal velocity field. Under these assumptions, the shallow water equations decouple if we consider the Bernoulli function as given. The bottom topography b will then become one of the unknowns.

We begin by defining the relative vorticity $\zeta = \nabla^\perp \cdot u \equiv \partial_1 u_2 - \partial_2 u_1$ and the Bernoulli function $B(x) = \frac{1}{2} |u(x)|^2 + \rho(x) + b(x)$, so that the shallow water momentum equations read

$$(1 + \zeta) u^\perp + \nabla B = 0 . \tag{20}$$

We then note that the transport stream function Ψ, defined via

$$\rho u = \nabla^\perp \Psi , \tag{21}$$

is in the steady state only a function of x, and constant along parcel trajectories. It is known that the *potential vorticity* $q = (1 + \zeta)/\rho$ is materially conserved. Under the steady-state assumption, this implies $q(x) = q(\Psi(x))$. Plugging (21) into (20), we find that $q(\Psi) \nabla\Psi = \nabla B$, so that

$$\frac{dB}{d\Psi} = q(\Psi) = \frac{1+\zeta}{\rho}. \tag{22}$$

In the special case that u is divergence free, we can introduce a second stream function ξ via

$$u = \nabla^\perp \xi. \tag{23}$$

Since Ψ is constant along parcel trajectories, $\Psi(x) = \Psi(\xi(x))$. Plugging (23) into (21) yields $\rho \nabla\xi = \nabla\Psi$, so that ρ as function of ξ satisfies $\rho(\xi) = d\Psi/d\xi$. Hence, (22) can be written as

$$\frac{dB}{d\xi} = 1 + \zeta = 1 + \Delta\xi. \tag{24}$$

We now specialize to the class of *cosine vortex solutions* by setting

$$\xi(x) = c \, (\cos x_1 + a \, \cos x_2), \tag{25}$$

where a and c are constants. Consequently, $\nabla^2 \xi = -\xi$ so that (24) reads $dB/d\xi = 1 - \xi$. Integrating this relation, we obtain

$$B(\xi) = -\frac{1}{2}\xi^2 + \xi + B_0. \tag{26}$$

Recalling that

$$B(\xi) = \frac{1}{2}|\nabla\xi|^2 + \rho(\xi) + b, \tag{27}$$

we can determine $\rho(x) = \rho(\xi(x))$ and a consistent bottom profile $b(x)$. Namely, combining (26) and (27), and using (25), we obtain

$$\rho(\xi(x)) + b(x) = -\frac{c^2}{2}(1 + a^2 + 2a \, \cos x_1 \, \cos x_2) + \xi(x) + B_0. \tag{28}$$

This equation can be consistently partitioned into two relations $\rho(\xi) = \xi + \rho_0$ and $b(x) = -a \, c^2 \cos x_1 \cos x_2 + b_0$, where the constant ρ_0 is chosen large enough to ensure that $\rho(x) > 0$. The cosine vortex with the choice of parameters used in our benchmarks is shown in Fig. 2. We remark that the $x_1 - x_2$ plane can be seen as the planar phase space for the dynamics of a single fluid parcel. In this view, $(0, -\pi)$ and $(-\pi, 0)$ and their periodizations are equilibrium points; the line segments $x_2 = \pm x_1 \pm \pi$ are heteroclinic connections of these equilibrium points.

Fig. 2 Stream function ξ and
stream lines of the cosine
vortex (25) with $a = 1$ and
$c = -1, \rho_0 = 2.5$, and
$b_0 = 1$

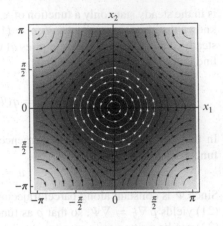

The vortex appears stable, although a stability analysis using the standard energy-Casimir method [15] is inconclusive. This shall be discussed in more detail in a forthcoming paper.

5 Results

In the following, we present a number of multi-parameter studies of the behavior of the HPM method for the two benchmark cases. To decouple the effects of the mesh size λ and the smoothing length μ as much as possible, we present the results in terms of the *relative smoothing*

$$\mu_{\text{rel}} = \mu/\Lambda, \tag{29}$$

where $\Lambda = 2\pi/L$ is the initial inter-particle distance. Thus, for $\mu_{\text{rel}} = 1$, the smoothing length scale and the inter-particle distance are comparable. We further let n denote the initial number of particles per cell per dimension, i.e.,

$$L = n K \qquad \text{and} \qquad N = L^d = n^d K^d. \tag{30}$$

As time integrator, we use the standard fourth order explicit Runge–Kutta scheme which is accurate and robust. In our simulations, time integration errors were consistently subdominant for time step $\tau = 10^{-4}$. As we are not working on long-time integration, there is no advantage in using a symplectic time integrator as in [6].

5.1 *Optimal Global Smoothing*

In the first series of benchmarks, we fix the initial number of particles per cell as this is the only regime where the computational effort of the grid operations and of the particle-to-grid operations remain asymptotically comparable (up to logarithms).

Fig. 3 Error levels for the cosine vortex with initially $n = 2$ particles per cell per dimension, $q = 6$, $p = 4$, and final time $T = 0.5$

Fig. 4 Error levels for the cosine vortex with initially $n = 1$ particle per cell per dimension, $q = 6$, $p = 4$, and final time $T = 0.5$

Fig. 5 Error levels for the cosine vortex with initially $n = 0.5$ particles per cell per dimension, $q = 6$, $p = 4$, and final time $T = 0.5$

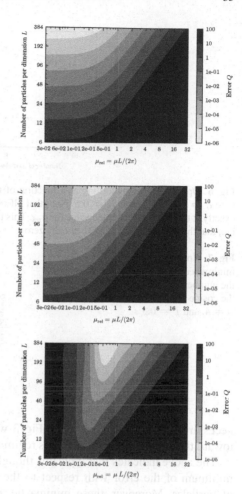

In the first set of tests, we also make sure that the sufficient conditions on the order of the Strang–Fix condition and on the order of the global smoothing under which Theorem 1 asserts convergence of the HPM method are satisfied. In Sects. 5.3 and 5.4, we later investigate what happens when these conditions are violated.

The qualitative behavior of the error under the change of the number of particles and of the global smoothing length is very similar in all cases. Figures 3–5 are examples for the cosine vortex; the behavior for the one-dimensional Burgers' solution is very similar. Two features are worth pointing out. First, when the global smoothing is too large, errors rise. In the regime of large global smoothing, contours of constant error have slope 1 in the μ_{rel}-L plot; however this is simply saying that $\mu = \text{const}$ and therefore to be trivially expected. Second, and more interestingly, when there is a sufficient number of particles per cell, lowest errors are obtained without global smoothing at all. With an increasing number of particles, however, global smoothing appears to become necessary to reduce errors, see Fig. 4. We do not know if, as the number of particles increases, global smoothing always becomes

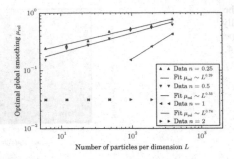

Fig. 6 Optimal global smoothing as a function of the number of particles for the cosine vortex with $q = 6$, $p = 4$, and $T = 0.5$. Note that all of the $n = 2$ and some of the $n = 1$ data lie exactly at the minimum sampled value for μ_{rel}. This indicates that global smoothing is not required, i.e., that the true optimal μ_{rel} is identically zero

Fig. 7 Corresponding minimal error as a function of the number of particles for the cosine vortex with $q = 6$, $p = 4$, and $T = 0.5$

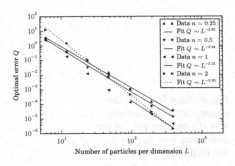

necessary, in other words, whether Fig. 3 will look like Fig. 4 if it were extended toward the larger number of particles regime.

The above indicates that typically, though not always, there is a unique nonzero minimum of the error with respect to the global smoothing for a fixed number of particles. Moreover, these minima lie approximately on a straight line when the coordinate axes are logarithmically scaled, implying that the optimal global smoothing satisfies a power law relationship with the number of particles.

Figure 6 shows the optimal global smoothing for four different initial numbers of particles per cell. The cases $n = 2$, $n = 1$, and $n = 0.5$ correspond directly to Figs. 3–5, respectively. For even more initial particles per cell, the behavior is very close to that of $n = 2$; for even fewer initial particles per cell, the behavior is very close to that of $n = 0.25$. The scaling of the corresponding optimal error is shown in Fig. 7.

5.2 Optimal Number of Particles per Cell

We can add another level of analysis to the preceding set of simulation data: suppose that for fixed values of the initial number of particles per dimension L and number of particles per cell per dimension n we always choose the optimal global smoothing

Fig. 8 Error levels for the optimally smoothed cosine vortex simulation with $q = 6$, $p = 4$, and final time $T = 0.5$

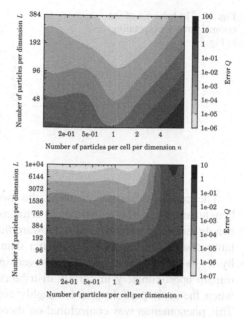

Fig. 9 Error levels for the optimally smoothed simulation of Burgers' solution with $q = 6$, $p = 4$, and final time $T = 0.95$

Fig. 10 Optimal choice of initial number of particles per cell per dimension n for the optimally smoothed HPM benchmarks shown in Figs. 8 and 9

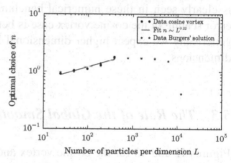

length scale. How then does the error behave as a function of L and n? And further, for given L, what is the optimal number of particles per cell?

If the cells are too coarse, then clearly we expect the local interpolation error to increase. Whether the error should increase in the opposite extreme, when we refine the mesh without adding more particles, is less obvious. Indeed, Fig. 8 shows that this is so for the cosine vortex benchmark, but that there is almost no distinct minimum of error for an intermediate number of particles per cell in the Burgers' benchmark, Fig. 9.

The optimal number of particles in the vortex case follows very roughly a power law, see Fig. 10. For Burgers' solution and a small number of particles, the scaling behavior is very similar, but Fig. 9 makes clear that in this case this behavior is very fragile and indeed breaks down when taking more particles. Yet, the optimal error follows a relatively stable power law in both cases, see Fig. 11.

Fig. 11 The behavior of the optimal error corresponding to Fig. 10

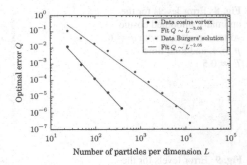

The crucial difference between the two cases is that in the cosine vortex benchmark, the particle distribution remains relatively uniform, while in Burgers' benchmark the particles bunch up toward the right hand side of the domain. In the latter case, HPM cannot do any better than SPH—the SPH limit is characterized by $n \to 0$ in Figs. 8 and 9. However, in the "good" case when the particles remain approximately uniformly distributed, HPM has a distinct error optimum when the scale particle grid is roughly comparable with that of the HPM grid. This phenomenon was conjectured on theoretical grounds in [11, Remark 1] and is clearly seen in these numerical benchmarks. It also explains why the optimal error scaling in the cosine vortex case is better than for Burgers' solution although generically we expect higher dimensional HPM to do worse than HPM in lower dimensions.

5.3   The Role of the Global Smoothing Order q

Figures 12 and 13 for the cosine vortex and Figs. 14 and 15 for Burgers' solution, respectively, show the influence of q on the optimal global smoothing length scale and the corresponding optimal rate of convergence. In the proof of Theorem 1 given in [11], a decay condition of the Fourier symbol of the smoothing kernel implies the requirement $q > p + d/2$. Numerically, we do not find any indication that convergence fails when this condition is violated. However, the value of q does feature significantly into the rate of convergence whenever the number of particles per cell drop below unity. (When $n \gg 1$, the optimal smoothing length is zero and the value of q does not enter the computation at all.)

The optimal error in the theoretical estimate is obtained by balancing the influence of the initialization error with that of a dynamical error contribution. This implies that the initialization error, which is essentially dominated by the smoothing error, should scale like the total observed error. Let us therefore look at the scaling of the pure smoothing error. To do so, we recall a well-known result regarding the L^p error for convolution smoothing on $\mathbb{R}^d$ [14]. Namely, provided the kernel function satisfies a first order moment condition and some further technical requirements,

Fig. 12 Optimal global
smoothing as a function of the
number of particles for the
cosine vortex with $n = 0.25$,
$p = 4$, and $T = 0.5$

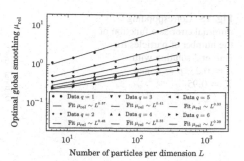

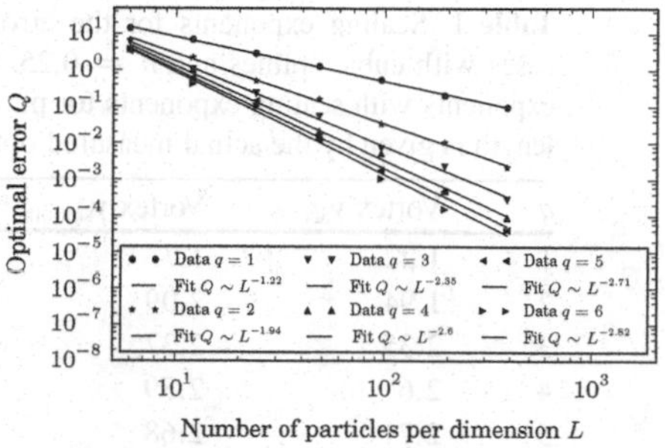

Fig. 13 Corresponding
minimal error as a function of
the number of particles for the
cosine vortex with $n = 0.25$,
$p = 4$, and $T = 0.5$

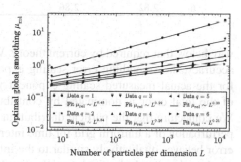

Fig. 14 Optimal global
smoothing as a function of
the number of particles for
Burgers' solution with
$n = 0.25$, $p = 4$, and
$T = 0.95$

the smoothing error is of second order in the smoothing length scale μ. Disregarding grid effects, this implies that whenever the optimal smoothing length has scaling $\mu \sim L^{-\kappa}$, the pure smoothing error contribution would scale like

$$Q_{\text{smooth}} \sim L^{-4\kappa} \equiv L^{-\gamma_{\text{smooth}}}. \qquad (31)$$

Obviously, the HPM dynamics and the grid approximation introduce many further contributions to the total error Q but, by comparing the respective scaling exponents, we shall be able to tell whether the smoothing error is dominant or subdominant.

The results of this analysis are displayed in Table 1, where the behavior for the two test cases is markedly different: for the cosine vortex, the smoothing error is the dominant error contribution except for $q = 1, 2$ when the order of the operator is

Fig. 15 Corresponding
minimal error as a function of
the number of particles for
Burgers' solution with
$n = 0.25$, $p = 4$, and
$T = 0.95$

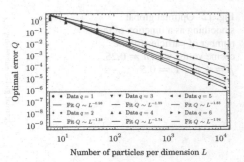

Table 1 Scaling exponents for the error so that $Q \sim L^{-\gamma}$ for the two test
cases with cubic splines and $n = 0.25$. Compared are actual measured scaling
exponents with scaling exponents for pure smoothing error where the smoothing
length is given by the actual measured optimal smoothing length

q	Vortex γ_{act}	Vortex γ_{smooth}	Burgers γ_{act}	Burgers γ_{smooth}
1	1.22	1.7	0.98	2.19
2	1.94	2.09	1.35	2.62
3	2.35	2.37	1.59	2.85
4	2.6	2.59	1.74	2.97
5	2.71	2.68	1.83	3.08
6	2.82	2.86	1.94	3.14

way below that required by current theory. We see clearly that, within the expected
accuracy, the smoothing error scaling exponent γ_{smooth} provides an upper bound
for the actual measured scaling exponent γ_{act}. For Burgers' solution, the scaling
of the total error is dominated by contributions other than the smoothing error.
We conjecture that the particle distribution for the cosine vortex can be seen as a
perturbation of a uniform grid for time intervals of order one so that the dynamical
error behaves qualitatively similar to the initialization error. For Burgers' solution,
on the other hand, the final particle distribution is very nonuniform, so that the
dynamical error is distinctly worse than the initialization error; therefore it must
be distinctly worse than pure smoothing error scaling. We further believe that the
two examples give two extremes of the behavior of HPM under the assumption that
the true solution is smooth, and that generic behavior should be within the range
covered by these examples. This clearly requires further investigation.

Let us also compare the numerical values of the observed to the theoretically
predicted scaling exponents (12) and (13) for cubic splines as local interpolation
kernels which, for $p = 4$, read

$$\mu \sim L^{-\frac{3}{6+d}} \quad \text{and} \quad Q \sim L^{-\frac{6}{6+d}}. \tag{32}$$

The comparison for $q = 6$ is summarized in Table 2. Clearly, the error bounds
given by Theorem 1 are far from sharp. In the case of the cosine vortex where the
error behavior is dominated by the smoothing error, the following interpretation is

Table 2 Scaling exponents for optimal smoothing $\mu \sim L^{-\kappa}$ and corresponding error $Q \sim L^{-\gamma}$ for the two test cases with cubic splines, $q = 6$, and $n = 0.25$. Notice that $\mu \sim \mu_{rel}/L$

	κ_{theory}	κ_{act}	γ_{theory}	γ_{act}
Vortex	0.38	0.71	0.75	2.82
Burgers	0.43	0.79	0.85	1.94

Fig. 16 Minimal error as a function of the number of particles for the cosine vortex with $n = 2$, $q = 6$, and $T = 0.5$. The values $p = 3, \ldots, 6$ correspond to quadratic, cubic, quartic, and quintic cardinal B-splines, respectively

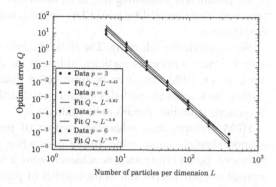

possible: The scaling exponent of the initialization error is underestimated in [11] by a factor of 2 due to the L^1-L^∞ splitting used there. As the dynamical error of the cosine vortex behaves qualitatively like the initialization error, this would put the optimal global smoothing scaling exponent off by a factor of 2, and then again the scaling relation between μ and Q off by a factor of 2, so that we might expect an underestimate by a factor of 4 altogether. The data is roughly consistent with this explanation, but a more in-depth investigation will be necessary to come to a firm conclusion. In the case of Burgers' solution, the underestimate is not as dramatic as the dominant error contributions are due to other effects which are presumably captured better by the theory.

5.4 The Role of the Strang–Fix Order p

To our surprise, the order of the Strang–Fix condition has very little influence on the performance of the HPM method, as Fig. 16 illustrates. The support of a linear spline is simply too small to ensure consistent approximation. However, in the quadratic case where $p = 3$, the approximation is already quite good with only a slight degradation for large numbers of particles. Moreover, there is no improvement going beyond cubic splines; indeed, the error even rises very slightly, which is likely due to the increased support of the higher order splines and the fact that the examples are taken in a regime where global smoothing for providing long-range interactions is not required.

6 Conclusion

We have achieved a comprehensive numerical study of the parameter dependence of the HPM method in a simple, controlled setting. Within the constraints of this setting (looking at the short-time error only, special solutions instead of generic data, and limited parameter ranges), we found that the error behavior as a function of the parameters smoothing length μ, mesh size λ, and number of particles N is very well characterized by power law dependencies, and thus can be described in simple terms.

We conclude the following. The HPM method performs at least as good as SPH with the same number of particles, although the asymptotic computational cost is much lower. HPM may even perform better than SPH with the same number of particles so long as the particles remain well distributed. Whether this is relevant in the generic situation remains to be studied.

HPM performs best when the number of particles and the number of cells is approximately comparable. In the case when the particle distribution remains favorable, best performance is achieved with a slight asymptotic increase of the number of particles per cell as the number of particles increases. This, however, is likely not relevant in practice, so that a constant number of particles per cell at, or slightly above unity, appears advisable.

The decay exponent q of the global smoothing kernel plays a big role in determining the observed rate of convergence, where larger powers yield better results, although improvements are slight beyond $q = 3$ or 4. However, we do not see a break-down of convergence for low values of q as might be expected from theory.

In contrast, the order of the local interpolation spline plays almost no role for the error behavior. We see a possible slight improvement going from spline order 2 to 3, but no further benefit beyond that. As the number of interacting grid nodes and the associated computational cost increases with the spline order, the cubic spline appears to be the best choice. Whether an optimally written code with quadratic splines might beat out the cubic version in terms of error per CPU time is beyond what we can assess presently.

We finally remark that our scalings are stated in terms of the number of particles. Since the computational complexity is log-linear in the number of particles, this correlates approximately with scaling in terms of CPU time. However, as the discussion of the spline order shows, the constant factors can differ appreciably.

In a forthcoming study, we shall look specifically into the role of the Hamiltonian structure for simulations over longer time scales, and into the behavior with generic data.

Acknowledgements V.M. was supported in part through German Science Foundation grant LI-1530/6-1. M.O. acknowledges support through German Science Foundation grant OL-155/5-1 and through the European Science Foundation network Harmonic and Complex Analysis and Applications (HCAA). B.P. was supported by the Netherlands Organisation for Scientific Research (NWO) under the grant "Hamiltonian-based numerical methods in forced-dissipative climate prediction".

References

1. V.R. Ambati, O. Bokhove, Space-time discontinuous Galerkin discretization of rotating shallow water equations. J. Comput. Phys. **225**, 1233–1261 (2007)
2. O. Bokhove, M. Oliver, Parcel Eulerian–Lagrangian fluid dynamics for rotating geophysical flows. Proc. R. Soc. A **462**, 2563–2573 (2006)
3. C.J. Cotter, J. Frank, S. Reich, Hamiltonian particle-mesh method for two-layer shallow-water equations subject to the rigid-lid approximation. SIAM J. Appl. Dyn. Syst. **3**, 69–83 (2004)
4. J. Frank, S. Reich, Conservation properties of smoothed particle hydrodynamics applied to the shallow water equations. BIT **43**, 40–54 (2003)
5. J. Frank, S. Reich, The hamiltonian particle-mesh method for the spherical shallow water equations. Atmos. Sci. Lett. **5**, 89–95 (2004)
6. J. Frank, G. Gottwald, S. Reich, A Hamiltonian particle-mesh method for the rotating shallow water equations, in *Meshfree Methods for Partial Differential Equations*, ed. by M. Griebel, M.A. Schweitzer. Lecture Notes in Computational Science and Engineering, vol. 26 (Springer, Berlin, 2002), pp. 131–142
7. R.A. Gingold, J.J. Monaghan, Smoothed particle hydrodynamics: theory and application to non-spherical stars. Mon. Not. R. Astron. Soc. **181**, 375–389 (1977)
8. R. Iacono, Analytic solution to the shallow water equations. Phys. Rev. E **72**, 017302 (2005)
9. L.B. Lucy, A numerical approach to the testing of the fission hypothesis. Astrophys. J. **82**, 1013 (1977)
10. V. Molchanov, Particle-mesh and meshless methods for a class of barotropic fluids. Ph.D. thesis, Jacobs University, 2008
11. V. Molchanov, M. Oliver, Convergence of the Hamiltonian particle-mesh method for barotropic fluid flow. Mat. Comp. published online, DOI: 10.1090/S0025-5718-2012-02648-2 (2012)
12. K. Oelschläger, On the connection between Hamiltonian many-particle systems and the hydrodynamical equations. Arch. Ration. Mech. Anal. **115**, 297–310 (1991)
13. D.J. Price, Smoothed particle hydrodynamics and magnetohydrodynamics. J. Comput. Phys. **231**, 759–794 (2012)
14. P. Raviart, An analysis of particle methods, in *Numerical Methods in Fluid Dynamics*, ed. by F. Brezzi. Lecture Notes in Mathematics, vol. 1127 (Springer, Berlin, 1985), pp. 243–324
15. P. Ripa, General stability conditions for a multi-layer model. J. Fluid. Mech. **222**, 119–137 (1991)
16. P. Tassi, O. Bokhove, C. Vionnet, Space-discontinuous Galerkin method for shallow water flows—kinetic and HLLC flux, and potential vorticity generation. Adv. Water Res. **30**, 998–1015 (2007)

Peridynamics: A Nonlocal Continuum Theory

Etienne Emmrich, Richard B. Lehoucq, and Dimitri Puhst

Abstract The peridynamic theory is a nonlocal theory of continuum mechanics based on an integro-differential equation without spatial derivatives, which can be easily applied in the vicinity of cracks, where discontinuities in the displacement field occur. In this paper we give a survey on important analytical and numerical results and applications of the peridynamic theory.

Keywords Peridynamics • Nonlocal model • Continuum mechanics

1 Introduction

Peridynamics is a nonlocal theory in continuum mechanics. Linear elastic behaviour of solids is well described by the partial differential equation

$$\rho(x)\partial_t^2 u(x,t) = (Lu)(x,t) + b(x,t), \quad (x,t) \in \Omega \times (0,T),$$
$$\text{with } (Lu)(x,t) := (\lambda + \mu)\,\text{grad}\,\text{div}\,u(x,t) + \mu\,\text{div}\,\text{grad}\,u(x,t), \tag{1}$$

which is derived from Newton's second law *Force = Mass × Acceleration*. In (1) ρ describes the density of the body; the right-hand side consists of the external force

E. Emmrich (✉) · D. Puhst
Institut für Mathematik, TU Berlin, Straße des 17. Juni 136, 10623 Berlin, Germany
e-mail: emmrich@math.tu-berlin.de; puhst@math.tu-berlin.de

R.B. Lehoucq
Multiphysics Simulation Technologies, Sandia National Laboratories, Albuquerque, NM 87185, USA
e-mail: rblehou@sandia.gov
Sandia is a multiprogram laboratory operated by Sandia Corporation, a Lockheed Martin Company, for the U.S. Department of Energy under contract DE-AC04-94AL85000.

M. Griebel and M.A. Schweitzer (eds.), *Meshfree Methods for Partial Differential Equations VI*, Lecture Notes in Computational Science and Engineering 89, DOI 10.1007/978-3-642-32979-1_3, © Springer-Verlag Berlin Heidelberg 2013

density b as well as inner tensions and macroscopic forces with Lamé parameters λ and μ. The variable $u : \bar{\Omega} \times [0, T] \to \mathbb{R}^d$ with $\Omega \subset \mathbb{R}^d$ and $d \in \{1, 2, 3\}$ is the displacement field. The limitation of the elastic model (1) is the implicit assumption that the deformation is twice continuously differentiable, which manifests in the inability to model spontaneous cracks and fractures. These are discontinuities of the displacement field and thus, (1) is not defined on cracks or fractures.

In 2000, Silling [42] proposed the peridynamic continuum theory that makes minimal regularity assumptions on the deformation. Instead of spatial differential operators, integration over differences of the displacement field is used to describe the existing, possibly nonlinear, forces between particles of the solid body. The resulting derivative-free nonlocal peridynamic equation of motion reads, in the general nonlinear form,

$$
\begin{aligned}
&\rho(x)\partial_t^2 u(x, t) \\
&= \int_{\mathcal{H}(x)} f(x, \hat{x}, u(x, t), u(\hat{x}, t), t)d\hat{x} + b(x, t), \quad (x, t) \in \Omega \times (0, T).
\end{aligned}
\tag{2}
$$

This theory is named after the greek words *peri* (near) and *dynamis* (force), because in (2) for every particle x the force exerted by each surrounding particle $\hat{x}$ on x is included. The integration domain $\mathcal{H}(x)$ describes the volume of particles interacting and is typically the open ball of radius δ surrounding x intersected with Ω. Then δ is called peridynamic horizon. Beyond this horizon we have

$$
f(x, \hat{x}, u(x, t), u(\hat{x}, t), t) = 0, \quad \hat{x} \notin \mathcal{H}(x). \tag{3}
$$

The reader is referred to the recent survey Silling and Lehoucq [48] emphasizing the mechanical aspects of the peridynamic continuum theory. This includes proposing the peridynamic balance of energy and thermodynamic restrictions so that the second law of thermodynamics is not violated. The principles of classical statistical mechanics are used to derive the energy and momentum balance laws in Lehoucq and Sears [34]. The aim of this contribution is to give a survey on important analytical and numerical results and applications of this nonlocal continuum theory.

Nonlocal theories taking into account effects of long-range interaction in elastic materials and their application to problems of solid and fracture mechanics have been studied for a long time, cf. the pioneering work by Kröner [32] and Eringen [28] and the references cited therein, the monographs by Kunin [33] and Rogula [40], and more recently (without being exhaustive) Altan [5, 6], Bažant and Jirásek [9], Chen et al. [15, 16], Lei et al. [35], Pisano and Fuschi [37], Polizzotto [38, 39], and Wang and Dhaliwal [49, 50]. In contrast to the peridynamic theory, all these theories rely on spatial derivatives.

This paper is organized as follows. In Sect. 2 we describe the peridynamic model. First, the bond-based model is demonstrated, where the focus lies on the pairwise force function f. Then we derive the linearized model by linearizing the pairwise force function. Afterwards, we describe the state-based peridynamic model, which is a generalization of the bond-based model. Finally in Sect. 2, other nonlocal models in elasticity are presented.

In Sect. 3 a survey on the mathematical analysis of the linear bond-based model, the nonlinear bond-based model and state-based model is given. In addition, a nonlocal vector calculus is presented, which is a useful tool for describing the state-based model.

In Sect. 4 peridynamics is studied as a multiscale method. First, results on the limit of vanishing nonlocality are stated, that is the limit of (2) as the peridynamic horizon δ goes to zero. Then the modeling of composite materials including the use of two-scale convergence is considered.

Section 5 deals with the numerical approximation of peridynamics by the quadrature formula method.

Finally in Sect. 6 a few notes on applications and numerical simulations are given.

2 Peridynamic Model

2.1 Bond-Based Model

Since there are no spatial derivatives, boundary conditions are not needed in general for the partial integro-differential equation (2) (although this depends on the singularity behaviour of the integral kernel and the functional analytic setting). Nevertheless, "boundary" conditions can be imposed by prescribing u in a strip along the boundary constraining the solution along a nonzero volume. Hence, (2) is complemented with the initial data

$$u(\cdot, 0) = u_0 \quad \text{and} \quad \partial_t u(\cdot, 0) = \dot{u}_0 . \tag{4}$$

The interaction between particles within the given horizon is called bond. Since bonds are defined pairwise, this results in a Poisson ratio of $\nu = 1/4$. Thus, due to the relationship $\lambda = 2\mu\nu/(1 - 2\nu)$, the Lamé parameters λ and μ are identical in the linearized bond-based peridynamic theory. Linear elastic, isotropic material can have a Poisson number between -1 and $1/2$ though many examples can already be found for a Poisson number close to $1/4$ such as composite materials, polymers, sand, glass, and iron.

In the following we will use the notation

$$\xi = \hat{x} - x, \quad \eta = u(\hat{x}, t) - u(x, t).$$

Note that $\xi + \eta$ is the relative position of the deformed configuration (see Fig. 1). Assuming no explicit dependence on time and dependence only on differences in the displacement field, the pairwise force function reads

$$f(x, \hat{x}, u(x, t), u(\hat{x}, t), t) = f(x, \hat{x}, \eta).$$

Fig. 1 Notation for
bond-based model

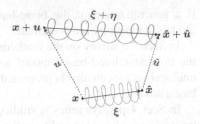

Combining Newton's third law *actio et reactio*, which gives

$$f(\hat{x}, x, -\eta) = -f(x, \hat{x}, \eta),$$ (5)

with the balance of angular momentum leads to the conclusion that the pairwise
force function is parallel to $\xi + \eta$. We call f homogenous if

$$f(x, \hat{x}, \eta) = f(\xi, \eta)$$

is fulfilled for all ξ and η.[1] Furthermore, a material is called microelastic if there
exists a so-called pairwise micropotential w such that $f(\xi, \eta) = \nabla_\eta w(\xi, \eta)$.
One of the simplest nonlinear models that has been suggested is the proportional
microelastic material model with

$$f(\xi, \eta) = c_{d,\delta}\, s(\xi, \eta)\, \frac{\xi + \eta}{|\xi + \eta|}\, \chi_{(0,\delta)}(|\xi|),$$ (6)

where

$$s(\xi, \eta) = \frac{|\xi + \eta| - |\xi|}{|\xi|}$$

denotes the bond stretch[2] that is the relative change of the length of a bond. By $|\cdot|$
we denote the Euclidean norm. In this case the related micropotential is given by

$$w(\xi, \eta) = \frac{c_{d,\delta}\, s^2(\xi, \eta)}{2}\, |\xi|\, \chi_{(0,\delta)}(|\xi|),$$

where we have chosen the micropotential such that $w(\xi, 0) = 0$. The constants of
proportionality are

$$c_{1,\delta} = \frac{18K}{5\delta^2}, \quad c_{2,\delta} = \frac{72K}{5\pi\delta^3}, \quad c_{3,\delta} = \frac{18K}{\pi\delta^4}$$

with the bulk modulus $K = \lambda(1 + v)/(3v) = 5\lambda/3$.

---

[1]Note that the notation $f = f(\xi, \eta)$ is somewhat ambiguous. Indeed, for a given function $u = u(x, t)$ the mathematical correct way to describe f is using the Nemytskii operator $F : u \mapsto Fu$
with $(Fu)(x, \hat{x}, t) = f(\hat{x} - x, u(\hat{x}, t) - u(x, t))$.

[2]For the notation $s = s(\xi, \eta)$, see also Footnote 1.

Let the kinetic and elastic energy density be defined as

$$e_{kin}(x,t) = \frac{1}{2}\rho(x)|\partial_t u(x,t)|^2,$$

$$e_{el}(x,t) = \frac{1}{2}\int_{\mathscr{H}(x)} w(\hat{x} - x, u(\hat{x},t) - u(x,t))d\hat{x},$$

and let $l = e_{kin} - e_{el} - e_{ext}$ denote the total energy density with $e_{ext}(x,t) = -b(x,t){\cdot}u(x,t)$ induced by the external force density b. Then Eq. (2) can be derived from the variational problem of minimizing the functional

$$u \mapsto \int_0^T \int_\Omega l(x,t)dx dt,$$

see also Emmrich and Weckner [26].

In order to model cracks and to simulate fracture, bonds are allowed to break. This is realized by multiplying the pairwise force function, given, e.g., by (6), with the function

$$\mu(\xi,\eta,t) = \begin{cases} 1 & \text{if } s(\xi, u(\hat{x},\tau) - u(x,\tau)) \le s_0 \ \forall \tau \le t \\ 0 & \text{otherwise} \end{cases}$$

for some critical bond stretch s_0 for breakage (see also Silling and Askari [45] as well as Parks et al. [36]). Note that the resulting pairwise force function now explicitly depends on time t.

2.2 Linearization

A first-order approximation justifies for small relative displacements η the general linear ansatz

$$f(\xi,\eta) = f_0(\xi) + C(\xi)\eta$$

with the stiffness tensor (or micromodulus function) $C = C(\xi)$ and f_0 denoting forces in the reference configuration. Without loss of generality, we may assume $f_0 \equiv 0$ since otherwise f_0 can be incorporated into the right-hand side b. In general the stiffness tensor C is not definite. However, C has to be symmetric with respect to its argument as well as with respect to its tensor structure such that $C(\xi) = C(-\xi)$ and $C(\xi)^T = C(\xi)$. The corresponding micropotential is then given by $w(\xi,\eta) = \eta^T C(\xi)\eta/2$. In view of (3) we shall require $C(\xi) = 0$ if $|\xi| \ge \delta$.

The stiffness tensor can be shown to read as

$$C(\xi) = \lambda_{d,\delta}(|\xi|)\xi \otimes \xi$$

for a linear microelastic material (see also Silling [42]), where $\otimes$ denotes the dyadic product. The function $\lambda_{d,\delta} : \mathbb{R}_0^+ \to \mathbb{R}$ with $\lambda_{d,\delta}(r) = 0$ for $r \geq \delta$ determines the specific material model and depends on the dimension d and the peridynamic horizon δ. The linear peridynamic equation of motion (2) now reads as

$$\rho(x)\partial_t^2 u(x,t) = (L_{d,\delta}u)(x,t) + b(x,t), \quad (x,t) \in \Omega \times (0,T),$$

with

$$(L_{d,\delta}u)(x,t) := \int_{\mathcal{H}(x)} \lambda_{d,\delta}(|\hat{x} - x|)(\hat{x} - x) \otimes (\hat{x} - x) \cdot (u(\hat{x},t) - u(x,t))d\hat{x}.$$

$$(7)$$

The recent report Du et al. [21] also demonstrates how the one-dimensional equation (7) can be written as two first-order in time nonlocal advection equations. Note that $\lambda_{d,\delta}$ can have a singularity at $r = 0$. The standard example is the linearization of (6) with $\lambda_{d,\delta}(r) = c_{d,\delta}r^{-3}\chi_{(0,\delta)}(r)$. Unfortunately, in this model, the interaction jumps to zero if $r = \delta$. This jump discontinuity can be avoided by taking $\lambda_{d,\delta}(r) = \tilde{c}_{d,\delta}r^{-3}\exp(-\delta^2/(\delta^2 - r^2))\chi_{(0,\delta)}(r)$ with a suitable constant of proportionality $\tilde{c}_{d,\delta}$, see also Emmrich and Weckner [25]. This is of advantage also for the numerical approximation relying on quadrature.

2.3 State-Based Model

The integral

$$\int_{\mathcal{H}(x)} f\left(\hat{x} - x, u(\hat{x},t) - u(x,t)\right) d\hat{x}$$

expresses that the internal force density at x is a summation of forces over all vectors $\hat{x} - x$. Moreover, the summands are independent from each other. As demonstrated by Silling [42, §11], this effectively limits Poisson's ratio to a value of one-fourth for homogenous deformations of linear isotropic solids. In the same paper, a generalization is presented that augments the above integral with the term

$$e(\vartheta(x)), \text{ where } \vartheta(x) = \int j(|\hat{x} - x|)|\hat{x} - x + u(\hat{x},t) - u(x,t)| d\hat{x}.$$

The quantity ϑ is a weighted average of all the bonds $\hat{x} - x + u(\hat{x},t) - u(x,t)$ and provides the volume of a deformed sphere that is centered at x in the reference configuration; the quantity e then acts as a volume-dependent strain energy term that incorporates the collective motion of all the $\hat{x} - x$ simultaneously. The linearized operator is then shown to circumvent the restriction of Poisson's ratio to a value of one-fourth to its allowable values (see Silling et al. [46]). The recent report

Seleson et al. [41] demonstrates a symbiotic relationship between this generalization and an embedded atom (EAM) molecular dynamic potential.

Silling et al. [46] introduces the peridynamic state theory, a subsequent generalization to the approach in Silling [42, §15]. The integral in (2) is replaced by $L_{PD}u$ where

$$(L_{PD}u)(x,t) = \int_{\mathscr{H}(x)} \Big(\underline{\mathbf{T}}[x,t]\langle \hat{x} - x\rangle - \underline{\mathbf{T}}[\hat{x},t]\langle x - \hat{x}\rangle\Big)\, d\hat{x} \qquad (8)$$

leading to the equation of motion

$$\rho(x)\partial_t^2 u(x,t) = (L_{PD}u)(x,t) + b(x,t), \ (x,t) \in \Omega \times (0,T). \qquad (9)$$

The term $L_{PD}u$ is a functional of displacement representing the internal force density (per unit volume) that is exerted upon x by other points in the body. The *force state* $\underline{\mathbf{T}}[x,t]$ at x at time t is a mapping from the bond $\hat{x} - x$ to a force density (per unit volume) at x and is assumed to be zero outside the horizon δ. A material model is therefore provided by a relation between the deformation near x and the force state at x, e.g.,

$$\underline{\mathbf{T}} = \underline{\mathbf{T}}(\underline{\mathbf{Y}}) \qquad (10)$$

where $\underline{\mathbf{Y}}$ is the *deformation state*, a mapping from bonds connected to any x to the deformed images of these bonds, e.g.,

$$\underline{\mathbf{Y}}[x,t]\langle \hat{x} - x\rangle = \hat{x} - x + u(\hat{x},t) - u(x,t). \qquad (11)$$

Although the deformation or force state maps vectors into vectors, they are more complex than second order tensors because the mapping may be nonlinear and even discontinuous. The constitutive relation (10) explains that the force state maps the collective motion about x in contrast to the relation $f(\hat{x} - x, u(\hat{x},t) - u(x,t))$. A bond-based force is recovered via the identification $\underline{\mathbf{T}}[x,t]\langle \hat{x} - x\rangle = \frac{1}{2}f(\hat{x} - x, u(\hat{x},t) - u(x,t))$ because then

$$\underline{\mathbf{T}}[x,t]\langle \hat{x} - x\rangle - \underline{\mathbf{T}}[\hat{x},t]\langle x - \hat{x}\rangle$$

$$= \frac{1}{2}f\big(\hat{x} - x, u(\hat{x},t) - u(x,t)\big) - \frac{1}{2}f\big(x - \hat{x}, u(x,t) - u(\hat{x},t)\big)$$

$$= f\big(\hat{x} - x, u(\hat{x},t) - u(x,t)\big)$$

where we invoked (5) for the last equality. This identification reveals another important distinction between the bond-based and state theories. Namely, that force interaction is carried by the bond in the former theory while the interaction is split between the force state at x and $\hat{x}$ in the latter theory.

The paper [46] introduces a linear peridynamic state material model generalizing the linear bond-based material model. Silling [44] further generalizes the linear peridynamic state material model. A key concept is that of the modulus state, a second order tensor,

$$\mathbb{K} = \nabla \underline{\mathbf{T}} \langle \underline{\mathbf{Y}} \rangle : \mathbb{R}^3 \times \mathbb{R}^3 \to \mathbb{R}^{3 \times 3},$$

the Fréchet derivative of the force state with respect to $\underline{\mathbf{Y}}$. The modulus state represents the peridynamic analogue of the fourth order tensor in the classical linear elastic theory and so allows us to linearize a force state about a time-independent deformation state $\underline{\mathbf{Y}}^0$, e.g.,

$$\underline{\mathbf{T}}(\underline{\mathbf{U}})\langle \hat{\mathbf{x}} - \mathbf{x} \rangle = \underline{\mathbf{T}}(\underline{\mathbf{Y}}^0)\langle \hat{\mathbf{x}} - \mathbf{x} \rangle + \int_{\mathscr{H}(x)} \mathbb{K}\langle \hat{\mathbf{x}} - \mathbf{x}, \tilde{\mathbf{x}} - \mathbf{x} \rangle \underline{\mathbf{U}}\langle \tilde{\mathbf{x}} - \mathbf{x} \rangle \, d\tilde{\mathbf{x}} \quad (12)$$

where $\underline{\mathbf{U}}$ is the displacement state field defined by

$$\underline{\mathbf{U}}[\mathbf{x}]\langle \hat{\mathbf{x}} - \mathbf{x} \rangle = \underline{\mathbf{Y}}^0[\mathbf{x}]\langle \hat{\mathbf{x}} - \mathbf{x} \rangle - (\hat{\mathbf{x}} - \mathbf{x}). \quad (13)$$

Assume that the body force $\mathbf{b}$ has the decomposition $\mathbf{b}^0 + \hat{\mathbf{b}}$ for some time-independent external force density $\mathbf{b}^0$ so that

$$\int_{\mathscr{H}(x)} \left(\underline{\mathbf{T}}^0[\mathbf{x}]\langle \hat{\mathbf{x}} - \mathbf{x} \rangle - \underline{\mathbf{T}}^0[\hat{\mathbf{x}}]\langle \mathbf{x} - \hat{\mathbf{x}} \rangle \right) d\hat{\mathbf{x}} + \mathbf{b}^0(\mathbf{x}) = 0, \quad (14)$$

where $\underline{\mathbf{T}}^0 \equiv \underline{\mathbf{T}}(\underline{\mathbf{Y}}^0)$. Inserting the linearized force state (12) into the equation of motion (9), and invoking relations (13) and (14) results in an equation of motion with body force $\hat{\mathbf{b}}$ and force state given by the integral operator of (12).

An important material model is the *elastic* peridynamic material, defined by

$$\underline{\mathbf{T}}(\underline{\mathbf{Y}}) = \nabla W(\underline{\mathbf{Y}}) \quad (15)$$

where ∇W is the Fréchet derivative of the scalar valued function W, the *strain energy density function*. This elastic material model is the peridynamic analogue of the classical elastic model

$$\sigma = \frac{\partial W}{\partial F},$$

where σ and F are the Piola–Kirchoff stress tensor and deformation gradient, respectively. The modulus state is then $\mathbb{K} = \nabla \nabla W(\underline{\mathbf{Y}})$. The reader is referred to Silling [44] for example material models.

2.4 Other Nonlocal Models in Elasticity Theory

In Sect. 1, references to various nonlocal models in elasticity are stated. Two examples are given in the following. In [22], Duruk et al. study the one-dimensional nonlocal nonlinear Cauchy problem

$$\partial_t^2 u(x,t) = \partial_x^2 \int_{\mathbb{R}} \beta(x - \hat{x})g(u(\hat{x},t))d\hat{x}, \quad (x,t) \in \mathbb{R} \times (0,\infty), \quad (16)$$

complemented with initial conditions. Here β is an integrable kernel function whose Fourier transform satisfies a certain growth condition, and g is some nonlinear function. Global existence and blow-up conditions are investigated. For a so-called double-exponential kernel β equation (16) turns out to be equivalent to the higher-order Boussinesq equation

$$u_{tt} - u_{xx} - au_{xxtt} + bu_{xxxxtt} = (g(u))_{xx}$$

with positive constants a, b. This example illustrates that, first, nonlocal models so far still depend on spatial derivatives and, second, nonlocal models may be related with local models described by partial differential equations with higher-order spatial derivatives.

In Eringen [28] and Polizzotto [38] the Eringen model is proposed and developed. Combining the equation of motion

$$\rho(x)\partial_t^2 u(x,t) = \operatorname{div} \sigma(u(x,t)) + b(x,t)$$

with a nonlocal formulation of the stress tensor

$$\sigma(v(x)) = \int_{\Omega} \alpha(|\hat{x} - x|)s(v(\hat{x}))d\hat{x},$$

which incorporates the entire body Ω with a suitable interaction kernel α and the classical stress s from Hooke's law

$$s(v) = S : \frac{1}{2}\left(\nabla v + (\nabla v)^T\right)$$

with stiffness tensor S, gives a nonlocal equation of motion in linear elasticity.

3 Mathematical Analysis of the Peridynamic Model

3.1 Linear Bond-Based Model in L^2

First results on existence, uniqueness and qualitative behaviour of solutions to the peridynamic equation of motion have been presented in Emmrich and Weckner [24] for the infinite peridynamic bar. Besides well-posedness in L^∞ also nonlinear dispersion relations as well as jump relations for discontinuous solutions have been studied.

In [23], Emmrich and Weckner have proved results on existence, uniqueness, and continuous dependence of the solution for the linear model in an L^p-setting for $p > 2$ if $d = 2$ and $p > 3/2$ if $d = 3$. Moreover, a formal representation of the exact solution and a priori estimates are given. In [25], Emmrich and Weckner have proved well-posedness of the linear model in $L^\infty(\Omega)^d$ and in $L^2(\Omega)^d$ under the condition

$$\int_0^\delta |\lambda_{d,\delta}(r)| \, r^{d+1} dr < \infty. \tag{17}$$

Theorem 1 ([25]). *If (17) is fulfilled, then there exists for every $u_0, \dot{u}_0 \in L^2(\Omega)^d$, $b \in L^1(0, T; L^2(\Omega)^d)$ a unique solution $u \in \mathscr{C}^1([0, T], L^2(\Omega)^d)$ to the initial-value problem (7) and (4) that satisfies the a priori estimate*

$$\|u\|_{\mathscr{C}^1([0,T],L^2(\Omega)^d)} \le C_{d,\delta} \left(\|u_0\|_{L^2(\Omega)^d} + \|\dot{u}_0\|_{L^2(\Omega)^d} + \|b\|_{L^1(0,T;L^2(\Omega)^d)} \right).$$

If $b \in \mathscr{C}([0, T], L^2(\Omega)^d)$ then $u \in \mathscr{C}^2([0, T], L^2(\Omega)^d)$.

Here, Ω denotes a bounded Lipschitz domain, and the constant $C_{d,\delta}$ depends on the dimension d and the horizon δ. Moreover, other properties of the peridynamic integral operator defined through (7) such as dissipativity and self-adjointness are analyzed in Emmrich and Weckner [25].

In [17, 51], Du and Zhou consider the case $\Omega = \mathbb{R}^d$. Let $\mathscr{M}_\lambda(\mathbb{R}^d)^d$ be the space of functions $u \in L^2(\mathbb{R}^d)^d$ with

$$\int_{\mathbb{R}^d} (\mathscr{F}u)(y) \cdot (I + M_\delta(y))(\mathscr{F}u) dy < \infty,$$

depending on $\lambda_{d,\delta}$ since

$$M_\delta(y) = \int_0^\delta \lambda_{d,\delta}(|\hat{x}|)(1 - \cos(y \cdot \hat{x}))\hat{x} \otimes \hat{x} \, d\hat{x}.$$

Here $\mathscr{F}u$ denotes the Fourier transform of u.

A natural condition coming from the comparison of the deformation energy density of a homogeneous body under isotropic expansion arising from the peridynamic model with the energy density known from the classical linear elasticity theory is

$$\int_0^\delta \lambda_{d,\delta}(r) r^{d+3} dr < \infty. \tag{18}$$

Theorem 2 ([17, 51]). *Assume $\lambda_{d,\delta}(r) > 0$ for $0 < r < \delta$, (18) and $u_0 \in \mathscr{M}_\lambda(\mathbb{R}^d)^d$, $\dot{u}_0 \in L^2(\mathbb{R}^d)^d$ and $b \in L^2(0, T; L^2(\mathbb{R}^d)^d)$. Then the initial-value problem (7) and (4) has a unique solution $u \in C([0, T], \mathscr{M}_\lambda(\mathbb{R}^d)^d)$ with $\partial_t u \in L^2(0, T; L^2(\mathbb{R}^d)^d)$.*

If in addition condition (17) is valid, then Du and Zhou show that the space $\mathcal{M}_\lambda(\mathbb{R}^d)^d$ is equivalent to the space $L^2(\mathbb{R}^d)^d$.

3.2   Linear Bond-Based Model in H^σ ($\sigma \in (0,1)$)

The solution of Theorem 2 can take values in a fractional Sobolev space. Indeed, if

$$c_1 r^{-2-d-2\sigma} \leq \lambda_{d,\delta}(r) \leq c_2 r^{-2-d-2\sigma} \quad \forall\, 0 < r \leq \delta \qquad (19)$$

for some exponent $\sigma \in (0,1)$ and positive constants c_1 and c_2, then Theorem 2 remains true and the space $\mathcal{M}_\lambda(\mathbb{R}^d)^d$ is equivalent to the fractional Sobolev space $H^\sigma(\mathbb{R}^d)^d$, see Du and Zhou [17,51].

Additionally also the stationary problem is investigated in [17,51].

3.3   Nonlinear Bond-Based Model

A first result towards the nonlinear model is Erbay et al. [27] analyzing the nonlinear elastic bar. They consider the one-dimensional initial value problem

$$\partial_t^2 u = \int_{\mathbb{R}} \alpha(\hat{x} - x) g(u(\hat{x},t) - u(x,t)) d\hat{x}, \quad x \in \mathbb{R},\, t > 0,$$
$$u(x,0) = u_0, \quad \partial_t u(x,0) = \dot{u}_0, \quad x \in \mathbb{R}. \qquad (20)$$

Applying Banach's fixed point theorem the following theorems are proven.

Theorem 3 ([27]). *Let $X = C_b(\mathbb{R})$ or let $X = L^p(\mathbb{R}) \cap L^\infty(\mathbb{R})$ with $1 \leq p \leq \infty$. Assume $\alpha \in L^1(\mathbb{R})$ and $g \in \mathscr{C}^1(\mathbb{R})$ with $g(0) = 0$. Then there exists $T > 0$ such that the Cauchy problem (20) is locally well-posed with solution in $\mathscr{C}^2([0,T], X)$ for initial data $u_0, \dot{u}_0 \in X$.*

Theorem 4 ([27]). *Let $X = \mathscr{C}_b^1(\mathbb{R})$ or let $X = W^{1,p}(\mathbb{R})$ with $1 \leq p \leq \infty$. Assume $\alpha \in L^1(\mathbb{R})$ and $g \in \mathscr{C}^2(\mathbb{R})$ with $g(0) = 0$. Then there exists $T > 0$ such that the Cauchy problem (20) is locally well-posed with solution in $\mathscr{C}^2([0,T], X)$ for initial data $u_0, \dot{u}_0 \in X$.*

The authors of [27] remark that the proofs of the above theorems can be easily adapted to the more general peridynamic equation with a nonlinear pairwise force function $f(\xi, \eta)$, where f is continuously differentiable in η for almost every ξ and fulfils additional assumptions. For a more specific type of nonlinearities, Erbay et al. [27] proved well-posedness in fractional Sobolev spaces.

Theorem 5 ([27]). *Let* $\sigma > 0$ *and* $u_0, \dot{u}_0 \in W^{\sigma,2}(\mathbb{R}) \cap L^\infty(\mathbb{R})$. *Assume* $\alpha \in L^1(\mathbb{R})$ *and* $g(\eta) = \eta^3$. *Then there exists* $T > 0$ *such that the Cauchy problem (20) is locally well-posed with solution in* $\mathscr{C}^2([0, T], W^{\sigma,2}(\mathbb{R}) \cap L^\infty(\mathbb{R}))$.

Furthermore, blow up conditions for these solutions are investigated, which we shall not present here.

3.4 State-Based Model and Nonlocal Vector Calculus

The recent reports Du et al. [18, 19] consider a nonlocal vector calculus building upon the ideas of Gunzburger and Lehoucq [29]. The nonlocal vector calculus is applied in Du et al. [20] to establish the well-posedness of the linear peridynamic state equilibrium equation. We briefly summarize the vector calculus and well-posedness for the case of a linear isotropic material.

Let Ω denote an open subset of $\mathbb{R}^d$ and let $\alpha = \alpha(x, y)$ denote an anti-symmetric mapping, i.e., $\alpha(x, y) = -\alpha(y, x)$. Define the nonlocal divergence operator of a tensor $\boldsymbol{\Psi}$ and its adjoint operator,

$$\mathscr{D}_t(\boldsymbol{\Psi})(x) = \int_{\mathbb{R}^d} \left(\boldsymbol{\Psi}(x, y) + \boldsymbol{\Psi}(y, x)\right) \cdot \alpha(x, y)\, dy \quad \text{for } x \in \Omega,$$

$$\mathscr{D}_t^*(v)(x, y) = -\left(v(y) - v(x)\right) \otimes \alpha(x, y) \quad \text{for } x, y \in \Omega,$$

established via $\left(\mathscr{D}_t(\boldsymbol{\Psi}), v\right)_\Omega = \left(\boldsymbol{\Psi}, \mathscr{D}_t^*(v)\right)_{\Omega \times \Omega}$, inner products on $L^2(\Omega)$ and $L^2(\Omega \times \Omega)$, respectively. The operators $\mathscr{D}_t(\boldsymbol{\Psi})$ and $\mathscr{D}_t^*(v)$ return a vector and tensor respectively. Let $\omega = \omega(x, y)$ and $\mathbf{U} = \mathbf{U}(x)$ denote a nonnegative scalar function and tensor, respectively. The choice $\boldsymbol{\Psi}(x, y) = \omega(x, y)\mathbf{U}(x)$ results in the *weighted nonlocal divergence operator* $\mathscr{D}_{t,\omega}$ *for tensors* and its adjoint operator,

$$\mathscr{D}_{t,\omega}(\mathbf{U})(x) = \mathscr{D}_t\left(\omega(x, y)\mathbf{U}(x)\right)(x) \quad \text{for } x \in \Omega,$$

$$\mathscr{D}_{t,\omega}^*(u)(x) = \int_{\mathbb{R}^d} \mathscr{D}_t^*(u)(x, y)\, \omega(x, y)\, dy \quad \text{for } x \in \Omega.$$

The distinction between $\mathscr{D}_t^*(v)$ and $\mathscr{D}_{t,\omega}^*(u)$ is that the former is a tensor on $\Omega \times \Omega$ while the latter is a tensor on Ω.

In the case of a linear peridynamic solid, define the operator

$$\mathscr{L}(u) = \mathscr{D}_t\left(\eta\left(\mathscr{D}_t^*(u)\right)^T\right) + \mathscr{D}_{t,w}\left(\gamma \operatorname{Tr}\left(\mathscr{D}_{t,\omega}^*(u)\right)\mathrm{I}\right) \tag{21}$$

where η and γ are materials constants. Mild conditions may also be supplied so that

$$\mathscr{L} \to L \equiv (\lambda + \mu)\operatorname{grad}\operatorname{div} + \mu \operatorname{div}\operatorname{grad}$$

in $H^{-1}(\mathbb{R}^d)$ as the horizon decreases. Here, μ and λ are the Lamé parameters that are given as expressions in terms of η and γ. The first term on the right-hand side of (21) corresponds to the bond-based material model and is the peridynamic analogue of μ div grad $+\mu$ grad div. The second term on the right-hand side of (21) accounts for volumetric terms associated with Poisson ratio not equal to one fourth and is the peridynamic analogue of λ grad div.

In words, the peridynamic operator $\mathscr{L}$ converges to the Navier operator as the horizon tends to zero where the horizon is given by the support of $\boldsymbol{\alpha} \cdot \boldsymbol{\alpha}$ and ω in the two operators in the right-hand side of (21). In contrast to the Navier operator, the peridynamic operator $\mathscr{L}$ only depends upon spatial differences in the displacement field and avoids an assumption of the weak differentiability. Hence discontinuities in the displacement field do not lead to an ill-posed system.

A minor simplification of the main result of Du et al. [20] is given by the following result associated with a linear isotropic peridynamic material.

Theorem 6 ([20]). *Let the nonzero volume $\Omega_c \subset \Omega$ with $\Omega_s \cup \Omega_c = \Omega$ and $\Omega_s \cap \Omega_c = \emptyset$ be given. The volume-constrained problem*

$$\begin{cases} \mathscr{L}(u)(x) = b(x), & x \in \Omega_s, \\ \\ u(x) = h_d(x), & x \in \Omega_c, \end{cases} \tag{22}$$

is well-posed on $L^2(\Omega_s)^d$ when $b \in L^2(\Omega_s)^d$, given the volume-constraint h_d and mild conditions on η and γ.

The volume-constraint is the nonlocal analogue of a boundary condition, necessary for the well-posedness of a boundary value problem. The above theorem is established by demonstrating that (22) is the optimality system associated with a quadratic functional with a unique minimizer. Nonlocal analogues of Korn's inequality and Green's identities are needed.

4 Peridynamics as a Multiscale Method

4.1 Limit of Vanishing Nonlocality

A fundamental question of the peridynamic theory was if it generalizes the conventional linear elastic theory. More precisely, does the nonlocal linear peridynamic equation of motion converge towards the Navier equation of linear elasticity? Indeed in [25], Emmrich and Weckner proved convergence in an interior subdomain under smoothness assumptions of the solution. Therefore, let $\Lambda \subset \mathbb{R}^+$ be a null sequence bounded from above by some $\delta_0 > 0$, Ω_0 be the interior subdomain defined as all $x \in \Omega$ such that dist$(x, \partial\Omega) > \delta_0$, $L_{d,\delta}$ the linear operator defined through (7) and L the operator corresponding to the Navier equation of linear elasticity defined through (1) (with $\lambda = \mu$ since $\nu = 1/4$).

Theorem 7 ([25]). *Let (17) be valid for all* $\delta \in \Lambda$ *and* $\lambda_{d,\delta}$ *be nonnegative. If* $v \in \mathscr{C}^2(\Omega)^d$ *then*

$$\|L_{d,\delta}v - Lv\|_{L^\infty(\Omega_0)^d} \to 0 \quad as \quad \delta \to 0 \quad (\delta \in \Lambda).$$

In addition, an expansion of $L_{d,\delta}v$ in terms of a series of differential operators of even order $2n$ ($n = 1, 2, \ldots$) applied to v can be shown for smooth v, where the second-order differential operator is the Navier operator and where the coefficients of the differential operators behave like $\delta^{2(n-1)}$. This resembles, in the one-dimensional case, the expansion in Arndt and Griebel [7] for the evolution equation of an atomistic system as well as the higher-order Boussinesq equation appearing in the work of Duruk et al. [22]. A similar observation has also been made in Emmrich and Weckner [24] when discussing the relation between the peridynamic elastic bar and nonlinear dispersion relations: different dispersion relations correspond to different micromodulus functions, which then correspond (for smooth solutions) to different higher-order differential equations.

Furthermore in [17], Du and Zhou have also investigated the limit of vanishing nonlocality in the case of the full space $\Omega = \mathbb{R}^d$ being then able to show convergence of the sequence of solutions.

Theorem 8 ([17]). *Let (18) be valid and* $\lambda_{d,\delta}(r) > 0$ *for* $0 < r < \delta$. *If* $u_0 \in H^1(\mathbb{R}^d)^d$, $\dot{u}_0 \in L^2(\mathbb{R}^d)^d$ *and* $b \in L^2(0,T;L^2(\mathbb{R}^d)^d)$, *then the solution of the initial-value problem (7) and (4) converges to the solution of the initial-value problem (1) and (4) as* $\delta \to 0$ *in the conventional norms of* $L^2(0,T;\mathscr{M}_\lambda(\mathbb{R}^d)^d) \cap H^1(0,T;L^2(\mathbb{R}^d)^d)$ *if*

$$\int_{B_\delta(0)} \lambda_{d,\delta}(|x|)\,|x|^4\,dx \to 2d(d+2)\mu \quad as \quad \delta \to 0.$$

Note that here no extra regularity of the solution is assumed. Recall that if (17) is also fulfilled, then $\mathscr{M}_\lambda(\mathbb{R}^d)^d = L^2(\mathbb{R}^d)^d$ and (18) is always fulfilled if the deformation energy density for isotropic expansion arising from the peridynamic model coincides with the energy density known from the classical linear elasticity theory.

Results for the nonlinear bond-based model are not known so far. The limit of vanishing nonlocality of the state-based model is investigated in Lehoucq and Silling [47], see also Sect. 3.4.

4.2 Composite Material and Two-Scale Convergence

In [4], Alali and Lipton focus on the modeling and analysis of heterogeneous media using the peridynamic formulation. Considering both macroscopic and microscopic dynamics in one model, the so-called two-scale convergence is essential to approximate a solution.

In detail, a peridynamic long-range force is perturbed by an oscillating short-range force representing the heterogeneities. For example, one can think of a fibre-reinforced composite. The deformations are assumed to be small. This justifies that both the long-scale macroscopic force function f_{long} and the short-scale microscopic force function f_{short} are linearizations of the bond-strech model (see Sect. 2.2). The connection between those two scales is described by a parameter ε. The displacement inside the composite is denoted by $u_\varepsilon = u_\varepsilon(x,t)$. The peridynamic equation of motion for heterogeneous material reads

$$
\rho_\varepsilon(x)\partial_t^2 u_\varepsilon(x,t) = \int_{\mathscr{H}_{long}(x)} f_{long}(\hat{x} - x, u_\varepsilon(\hat{x},t) - u_\varepsilon(x,t))d\hat{x}
$$
$$
+ \int_{\mathscr{H}_{short}(x)} f_{short}(x, \hat{x} - x, u_\varepsilon(\hat{x},t) - u_\varepsilon(x,t))d\hat{x} \qquad (23)
$$
$$
+ b_\varepsilon(x,t), \quad (x,t) \in \Omega \times (0,T).
$$

Note that the microscopic force function now explicitly depends on x since the bond strength is influenced by the distance of the *fast* parameters of x/ε and $\hat{x}/\varepsilon$. The macroscopic integration volume $\mathscr{H}_{long}(x)$ is the open ball with centre x and radius $\delta_1 > 0$ intersected with Ω, the microscopic integration volume $\mathscr{H}_{short}(x)$ is the open ball with centre x and radius $\varepsilon\delta_2 > 0$ intersected with Ω. Alali and Lipton [4] show well-posedness of (23) complemented with initial conditions. Furthermore they show that the displacement $u_\varepsilon = u_\varepsilon(x,t)$ can be approximated by a function $u = u(x, x/\varepsilon, t)$. Indeed, setting $y = x/\varepsilon, u = u(x, y, t)$ is the two-scale limit of $u_\varepsilon = u_\varepsilon(x,t)$, that is

$$
\int_{\Omega \times (0,T)} u_\varepsilon(x,t)\Psi(x, x/\varepsilon, t)\,dxdt \rightarrow \int_{\Omega \times Y \times (0,T)} u(x, y, t)\Psi(x, y, t)\,dx\,dy\,dt
$$

for all $\Psi \in C_c^\infty(\mathbb{R}^3 \times Y \times \mathbb{R}^+)$ that are Y-periodic in y, as ε tends to zero. Here $Y \subset \mathbb{R}^3$ is the unit period cube with the origin at the center. The function $u = u(x, y, t)$ is obtained from a partial integro-differential initial value problem, which is similar to (23) but independent of ε.

5 Numerical Approximation

In what follows, we only describe the quadrature formula method. For a finite element approximation, we refer to Chen and Gunzburger [14].

Let $\Omega \subset \mathbb{R}^d$ ($d \in \{2,3\}$) be (for simplicity) a polyhedral domain and let $\mathbb{T} = \{\mathscr{T}_\ell\}_{\ell \in \mathbb{N}}$ be a sequence of partitions of Ω such that

1. $\mathscr{T} = \{T_i\}_{i=1}^M$ for any $\mathscr{T} \in \mathbb{T}$ with T_i ($i = 1, \ldots, M$, $M \in \mathbb{N}$) being a closed triangle (tetrahedron) or parallelogram (parallelepiped) with nonempty interior $\mathring{T}_i$ such that

$$\bigcup_{i=1}^{M} T_i = \overline{\Omega}, \ \mathring{T}_i \cap \mathring{T}_j = \emptyset \text{ for } i \neq j \ (i, j = 1, \dots, M);$$

2. There exist constants $c_1 > 0$ and $c_2 > 0$ such that for all $\mathscr{T} \in \mathbb{T}$ and all $T \in \mathscr{T}$

$$h_{\max} := \max_{T \in \mathscr{T}} h_T \leq c_1 h_T \leq c_2 \rho_{\min} := \min_{T \in \mathscr{T}} \rho_T$$

with $h_T := \text{diam } T$ being the diameter of T and ρ_T being the radius of the largest inscribed ball of T (quasi-uniformity);

3. $h_{\max}(\mathscr{T}_\ell) := \max_{T \in \mathscr{T}_\ell} h_T \to 0$ as $\ell \to \infty$.

We now describe the quadrature formula method for the spatial approximation of (7). Let $\phi \in \mathscr{C}(\overline{\Omega})$. On each $T \in \mathscr{T} \in \mathbb{T}$, we consider the quadrature formula

$$Q_T[\phi] := \sum_{\mu=1}^{m_T} \omega_{T,\mu} \phi(\mathbf{x}_{T,\mu}) \approx \int_T \phi(\hat{\mathbf{x}}) d\hat{\mathbf{x}} =: I_T[\phi]$$

with the quadrature points $\mathbf{x}_{T,\mu} \in T$ (remember that T is closed and quadrature points can also lie on the boundary of T) and weights $\omega_{T,\mu} > 0 \ (\mu = 1, \dots, m_T \in \mathbb{N})$. We then get the composed quadrature formula

$$Q_\Omega[\phi] := \sum_{T \in \mathscr{T}} Q_T[\phi] \approx I_\Omega[\phi].$$

By introducing a global numbering of the quadrature points $\mathbf{x}_{T,\mu}$ (each quadrature point is only counted once), we find

$$Q_\Omega[\phi] = \sum_{j=1}^{N} \omega_j \phi(\mathbf{x}_j)$$

with globally numbered weights ω_j. Note that ω_j is the sum of all the local weights corresponding to the quadrature point $\mathbf{x}_j$. For having a consistent quadrature, certain additional standard assumptions are required, which we are not going to specify here.

Using the quadrature points as collocation points for the approximation of (7), we arrive at

$$\rho(\mathbf{x}_i) \partial_t^2 \mathbf{u}_i(t) = (L_{d,\delta}^{(h)} \mathbf{u}^{(h)})(\mathbf{x}_i, t) + \mathbf{b}(\mathbf{x}_i, t), \quad i = 1, \dots, N,$$

with $(L_{d,\delta}^{(h)} \mathbf{u}^{(h)})(\mathbf{x}_i, t)$

$$:= \sum_{j \in J(i)} \omega_j \lambda_{d,\delta}(|\mathbf{x}_j - \mathbf{x}_i|)(\mathbf{x}_j - \mathbf{x}_i) \otimes (\mathbf{x}_j - \mathbf{x}_i)(\mathbf{u}_j(t) - \mathbf{u}_i(t))$$

and with $\mathbf{u}_i(t)$ being an approximation of $\mathbf{u}(\mathbf{x}_i, t)$ and $\mathbf{v}^{(h)}(\mathbf{x}_i) := \mathbf{v}_i$ for a grid function $\mathbf{v}^{(h)}$. Since the integration only takes place in the ball of radius δ intersected with Ω (see also Fig. 2), we take

Fig. 2 Discretization of the
integration volume

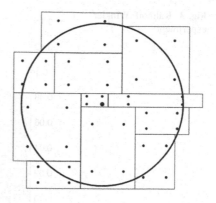

$$J(i) := \{ j \in \{1, 2, \ldots, N\} \setminus \{i\} : |x_j - x_i| < \delta \}.$$

As is seen, we take the global weight corresponding to a quadrature point even
if not all the elements to which this quadrature point belongs are fully inside the
peridynamic horizon. Moreover, if a quadrature point lies on the boundary of the
peridynamic horizon than it is not taken into account. One may object that it should
be better not to work with the global but with the local weight here. However, this
cannot influence the order of approximation.

Let us assume that the displacement field is smooth and that the local quadrature
rules are exact at least for polynomials of degree $r - 1$. Moreover, let us consider
the case of the linearization of the bond-based peridynamic model with a suitable
regularization in order to avoid the jump discontinuity along the boundary of the
peridynamic horizon such as $\lambda_{d,\delta}(r) = \tilde{c}_{d,\delta} r^{-3} \exp(-\delta^2/(\delta^2 - r^2)) \chi_{(0,\delta)}(r)$ (see
Emmrich and Weckner [25]).

Based upon an estimate of the local error of the quadrature rule, taking into
account the growth behaviour of the weak singularity of the integral kernel, and
employing stability of the solution with respect to perturbations of the integral
kernel, an error estimate of order $\mathcal{O}(h_{\max}^{\min(r,d)})$ and $\mathcal{O}(h_{\max}^d \log 1/h)$ if $r = d$,
respectively, has been shown in Büsing [13], where d denotes the dimension of the
domain. The restriction of the order to the dimension d could not be overcome
due to the singularity behaviour of the integral kernel.

6 Applications and Numerical Simulations

6.1 Simulation of Nanofibres

Investigating the behaviour of nanofibre networks and thin membranes of nanofibres
is of great economical and scientific interest. Bobaru [10] and Bobaru et al. [12]
have analyzed the behaviour of these fabrics under dynamic deformation including

Fig. 3 Kalthoff–Winkler
experiment

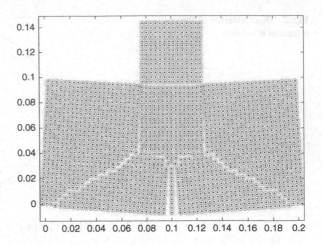

the influence of van der Waals forces. Assuming linear-elastic behaviour on each
fibre, we have a micropotential as given in Sect. 2.2. In order to avoid modeling
atoms, the van der Waals forces acting between fibres are modeled by adding the
so-called Lennard–Jones type potential of the form

$$w^{LJ}(|\xi + \eta|) = \alpha \left(\frac{a}{|\xi + \eta|} \right)^{12} - \beta \left(\frac{a}{|\xi + \eta|} \right)^{6}$$

with parametres α, β and a to the microelastic potential. Since van der Waals
forces are long-range forces, the linearized peridynamic equation of motion (7) is
complemented with an integral of the Lennard–Jones force density over a horizon,
which is different to δ.

6.2 Simulation of Cracks

Accurate crack prediction is of topical interest and many authors have worked on
it over the last years, see, e.g., Aidun and Silling [3], Agwai et al. [1, 2], Askari
et al. [8], Bobaru and Ha [11], Huang et al. [30], and Kilic and Madenci [31].
According to [3], the so-called Kalthoff–Winkler experiment is one of 12 characteristic phenomena of dynamic fracture, which a good numerical simulation should be
able to reproduce. In this experiment, a cylindrical steel impactor strikes on the
edge of a plate, which has been primed with two parallel notches. If the strike
is fast enough, the crack propagates in a characteristic angle (see Fig. 3 for a
simulation in Matlab with the quadrature formula method as described above and
a standard Runge–Kutta solver for the time integration). Silling [43] shows that the
Kalthoff–Winkler experiment is correctly simulated with the peridynamic theory.

In [3] the simulation of further characteristic phenomena of dynamic fracture with the peridynamic theory is investigated.

Acknowledgements The authors are grateful to Stephan Kusche and Henrik Büsing for the numerical simulation of the Kalthoff–Winkler experiment (see Fig. 3).

References

1. A. Agwai, I. Guven, E. Madenci, Peridynamic theory for impact damage prediction and propagation in electronic packages due to drop, in *Proceedings of the 58th Electronic Components and Technology Conference*, Lake Buena Vista, Florida (2008), pp. 1048–1053
2. A. Agwai, I. Guven, E. Madenci, Predicting crack propagation with peridynamics: a comparative study. Int. J. Fract. **171**(1), 65–78 (2011)
3. J.B. Aidun, S.A. Silling, Accurate prediction of dynamic fracture with peridynamics, in *Joint US-Russian Conference on Advances in Materials Science*, Prague (2009)
4. B. Alali, R. Lipton, Multiscale dynamics of heterogeneous media in the peridynamic formulation. J. Elast. **106**(1), 71–103 (2012)
5. B.S. Altan, Uniqueness of initial-boundary value problems in nonlocal elasticity. Int. J. Solid Struct. **25**(11), 1271–1278 (1989)
6. B.S. Altan, Uniqueness in nonlocal thermoelasticity. J. Therm. Stresses **14**, 121–128 (1991)
7. M. Arndt, M. Griebel, Derivation of higher order gradient continuum models from atomistic models for crystalline solids. Multiscale Model. Simul. **4**(2), 531–562 (2005)
8. A. Askari, J. Xu, S.A. Silling, Peridynamic analysis of damage and failure in composites, in *44th AIAA Aerospace Sciences Meeting and Exhibit*, Reno, AIAA-2006-88 (2006)
9. Z.P. Bažant, M. Jirásek, Nonlocal integral formulations of plasticity and damage: survey of progress. J. Eng. Mech. **128**(11), 1119–1149 (2002)
10. F. Bobaru, Influence of van der Waals forces on increasing the strength and toughness in dynamic fracture of nanofiber networks: a peridynamic approach. Model. Simul. Mater. Sci. **15**, 397–417 (2007)
11. F. Bobaru, Y.D. Ha, Studies of dynamic crack propagation and crack branching with peridynamics. Int. J. Fract. **162**(1–2), 229–244 (2010)
12. F. Bobaru, H. Jiang, S.A. Silling, Peridynamic fracture and damage modeling of membranes and nanofiber networks, in *Proceedings of the XI International Conference on Fracture*, Turin, vol. 5748 (2005), pp. 1–6
13. H. Büsing, *Multivariate Integration und Anwendungen in der Peridynamik*, Diploma thesis, TU Berlin, Institut für Mathematik, 2008
14. X. Chen, M. Gunzburger, Continuous and discontinuous finite element methods for a peridynamics model of mechanics. Comput. Method Appl. Mech. Eng. **200**(9–12), 1237–1250 (2011)
15. Y. Chen, J.D. Lee, A. Eskandarian, Dynamic meshless method applied to nonlocal crack problems. Theor. Appl. Fract. Mech. **38**, 293–300 (2002)
16. Y. Chen, J.D. Lee, A. Eskandarian, Atomistic viewpoint of the applicability of microcontinuum theories. Int. J. Solid Struct. **41**(8), 2085–2097 (2004)
17. Q. Du, K. Zhou, Mathematical analysis for the peridynamic nonlocal continuum theory. M2AN Math. Mod. Numer. Anal. **45**(2), 217–234 (2011)
18. Q. Du, M. Gunzburger, R.B. Lehoucq, K. Zhou, A nonlocal vector calculus, nonlocal volume-constrained problems, and nonlocal balance laws. Technical report 2010-8353J, Sandia National Laboratories, 2010
19. Q. Du, M. Gunzburger, R.B. Lehoucq, K. Zhou, Analysis and approximation of nonlocal diffusion problems with volume constraints. Technical report 2011-3168J, Sandia National Laboratories, 2011. Accepted for publication in SIAM review

20. Q. Du, M. Gunzburger, R.B. Lehoucq, K. Zhou, Application of a nonlocal vector calculus to the analysis of linear peridynamic materials. Technical report 2011-3870J, Sandia National Laboratories, 2011
21. Q. Du, J.R. Kamm, R.B. Lehoucq, M.L. Parks, A new approach for a nonlocal, nonlinear conservation law. SIAM J. Appl. Math. **72**(1), 464–487 (2012)
22. N. Duruk, H.A. Erbay, A. Erkip, Global existence and blow-up for a class of nonlocal nonlinear Cauchy problems arising in elasticity. Nonlinearity **23**, 107–118 (2010)
23. E. Emmrich, O. Weckner, The peridynamic equation of motion in non-local elasticity theory, in *III European Conference on Computational Mechanics. Solids, Structures and Coupled Problems in Engineering*, ed. by C.A. Mota Soares et al. (Springer, Lisbon, 2006)
24. E. Emmrich, O. Weckner, Analysis and numerical approximation of an integro-differential equation modelling non-local effects in linear elasticity. Math. Mech. Solid **12**(4), 363–384 (2007)
25. E. Emmrich, O. Weckner, On the well-posedness of the linear peridynamic model and its convergence towards the Navier equation of linear elasticity. Commun. Math. Sci. **5**(4), 851–864 (2007)
26. E. Emmrich, O. Weckner, The peridynamic equation and its spatial discretisation. Math. Model. Anal. **12**(1), 17–27 (2007)
27. H.A. Erbay, A. Erkip, G.M. Muslu, The Cauchy problem for a one dimensional nonlinear elastic peridynamic model. J. Differ. Equ. **252**, 4392–4409 (2012)
28. A.C. Eringen, Vistas of nonlocal continuum physics. Int. J. Eng. Sci. **30**(10), 1551–1565 (1992)
29. M. Gunzburger, R.B. Lehoucq, A nonlocal vector calculus with application to nonlocal boundary value problems. Multscale Model. Simul. **8**, 1581–1598 (2010). doi:10.1137/090766607
30. D. Huang, Q. Zhang, P. Qiao, Damage and progressive failure of concrete structures using non-local peridynamic modeling. Sci. China Technol. Sci. **54**(3), 591–596 (2011)
31. B. Kilic, E. Madenci, Prediction of crack paths in a quenched glass plate by using peridynamic theory. Int. J. Fract. **156**(2), 165–177 (2009)
32. E. Kröner, Elasticity theory of materials with long range cohesive forces. Int. J. Solid Struct. **3**, 731–742 (1967)
33. I.A. Kunin, *Elastic Media with Microstructure*, vol. I and II (Springer, Berlin, 1982/1983)
34. R.B. Lehoucq, M.P. Sears, Statistical mechanical foundation of the peridynamic nonlocal continuum theory: energy and momentum conservation laws. Phys. Rev. E **84**, 031112 (2011)
35. Y. Lei, M.I. Friswell, S. Adhikari, A Galerkin method for distributed systems with non-local damping. Int. J. Solid Struct. **43**(11–12), 3381–3400 (2006)
36. M.L. Parks, R.B. Lehoucq, S.J. Plimpton, S.A. Silling, Implementing peridynamics within a molecular dynamics code. Comput. Phys. Commun. **179**(11), 777–783 (2008)
37. A.A. Pisano, P. Fuschi, Closed form solution for a nonlocal elastic bar in tension. Int. J. Solid Struct. **40**(1), 13–23 (2003)
38. C. Polizzotto, Nonlocal elasticity and related variational principles. Int. J. Solid Struct. **38** (42–43), 7359–7380 (2001)
39. C. Polizzotto, Unified thermodynamic framework for nonlocal/gradient continuum mechanics. Eur. J. Mech. A Solid **22**, 651–668 (2003)
40. D. Rugola, *Nonlocal Theory of Material Media* (Springer, Berlin, 1982)
41. P. Seleson, M.L. Parks, M. Gunzburger, Peridynamic solid mechanics and the embedded atom model. Technical report SAND2010-8547J, Sandia National Laboratories, 2011
42. S.A. Silling, Reformulation of elasticity theory for discontinuities and long-range forces. J. Mech. Phys. Solid **48**(1), 175–209 (2000)
43. S.A. Silling, *Peridynamic Modeling of the Kalthoff–Winkler Experiment*. Submission for the 2001 Sandia Prize in Computational Science, 2002
44. S.A. Silling, Linearized theory of peridynamic states. J. Elast. **99**, 85–111 (2010)
45. S.A. Silling, E. Askari, A meshfree method based on the peridynamic model of solid mechanics. Comput. Struct. **83**, 1526–1535 (2005)
46. S.A. Silling, M. Epton, O. Weckner, J. Xu, E. Askari, Peridynamic states and constitutive modeling. J. Elast. **88**, 151–184 (2007)

47. S.A. Silling, R.B. Lehoucq, Convergence of peridynamics to classical elasticity theory. J. Elast. **93**(1), 13–37 (2008)
48. S.A. Silling, R.B. Lehoucq, Peridynamic theory of solid mechanics. Adv. Appl. Mech. **44**, 73–168 (2010)
49. J. Wang, R.S. Dhaliwal, On some theorems in the nonlocal theory of micropolar elasticity. Int. J. Solid Struct. **30**(10), 1331–1338 (1993)
50. J. Wang, R.S. Dhaliwal, Uniqueness in generalized nonlocal thermoelasticity. J. Therm. Stresses **16**, 71–77 (1993)
51. K. Zhou, Q. Du, Mathematical and numerical analysis of linear peridynamic models with nonlocal boundary conditions. SIAM J. Numer. Anal. **48**(5), 1759–1780 (2010)

Immersed Molecular Electrokinetic Finite Element Method for Nano-devices in Biotechnology and Gene Delivery

Wing Kam Liu, Adrian M. Kopacz, Tae-Rin Lee, Hansung Kim, and Paolo Decuzzi

Abstract It has been demonstrated from recent research that modern uses of multiscale analysis, uncertainty quantification techniques, and validation experiments is essential for the design of nanodevices in biotechnology and medicine. The 3D immersed molecular electrokinetic finite element method (IMEFEM) will be presented for the modeling of micro fluidic electrokinetic assembly of nanowires, filaments and bio-molecules. This transformative bio-nanotechnology is being developed to enable gene delivery systems to achieve desired therapeutic effects and for the design and optimization of an electric field enabled nanotip DNA sensor. For the nanodiamond-based drug delivery device we will discuss the multiscale analysis, quantum and molecular mechanics, immersed molecular finite element and meshfree methods, uncertainty quantification, and validation experiments. In addition, we will describe the mathematical formulation of pH control interactions among chemically functionalized nanodiamonds, and their interactions with polymers. For the nanotip sensor, we will discuss the underlying mechanics and physical parameters influencing the bio-sensing efficiency, such as the threshold of applied electric field, biomolecule deformation, and nanoscale

W.K. Liu (✉)
Walter P. Murphy Professor, Department of Mechanical Engineering, Northwestern University, Evanston, IL, USA

Visiting Distinguished Chair Professor, School of Mechanical Engineering, World Class University (WCU) Program in Sungkyunkwan University, Suwon, Korea
e-mail: w-liu@northwestern.edu

A.M. Kopacz · T.-R. Lee · H. Kim
Department of Mechanical Engineering, Northwestern University, Evanston, IL, USA
e-mail: a-kopacz@northwestern.edu; taerin-lee@northwestern.edu; han-kim@northwestern.edu

P. Decuzzi
The Methodist Hospital Research Institute, Department of Nanomedicine and Biomedical Engineering, Houston, TX, USA
e-mail: pdecuzzi@tmhs.org

M. Griebel and M.A. Schweitzer (eds.), *Meshfree Methods for Partial Differential Equations VI*, Lecture Notes in Computational Science and Engineering 89, DOI 10.1007/978-3-642-32979-1_4, © Springer-Verlag Berlin Heidelberg 2013

Brownian motion. Through multiscale analysis, we provide guidelines for nanodevice design, including fundamental mechanisms driving the system performance and optimization of distinct parameters.

Keywords IMEFEM • Sensors • Nanoparticles • Nanodiamonds

1 Nanotip Based DNA Preconcentration

One of the challenges in the fields of medicine is to develop a sensor that can detect biomolecules at ultra low concentrations. In order to identify such biological agents in a patient's blood or other bodily fluids at the onset of infection, a time-dependent dielectrophoretic force driven sensor is being developed. The immersed molecular electrokinetic finite element method (IMEFEM) framework is utilized to help guide the design of the device and characterize its fundamental mechanisms at the nanometer scale [3].

The immersed molecular electrokinetic finite element method is utilized to examine the concentration mechanism of the nanotip. Originally, the immersed finite element method (IFEM) has been developed by integrating the concept of immersed boundary, finite element and meshfree methods [10, 14, 15]. IFEM was later extended to incorporate electrohydrodynamics, namely IEFEM, and recently further enhancements incorporated molecular interactions and dynamics IMFEM [4, 12]. The immersed molecular electrokinetic finite element method coupled with multiphysics features were utilized to model dielectrophoresis, Brownian motion, and DNA particle-particle interactions [2, 6–9, 11].

A device composed of a 500 nm in diameter nanotip immersed in a fluid medium surrounded by a coil is shown in Fig. 1. The initial analysis of the system consisted of solving the governing equations of the electric field where the inherent carriers are the major current carriers. A high AC frequency of 5 MHz was applied across the nanotip and coil electrodes with a peak-to-peak voltage (Vpp) of 20 V. As shown in Fig. 2, an oligomer is place around the nanotip and considered as an elastic material in our calculation. The dielectrophoresis (DEP) force is the dominated force for attracting the oligomer to the nanotip. For the larger distance, the Brownian motion forces due to thermal motion are the dominating forces and the effectiveness of the DEP force exerted on an oligomer is diminished.

2 Vascular Nanoparticle Transport

Nanoparticles are good candidates to deliver drugs to target area where damaged cells or diseases are existed in. First of all, the nanoparticles for targeting the specific region are injected into blood stream. Secondly, they have to overcome the complex environment of blood flow to arrive at the estimated destination. Finally,

Fig. 1 Nanotip sensor device

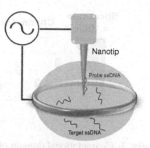

Fig. 2 DNA preconcentration at the nanotip using the electrokinetic force

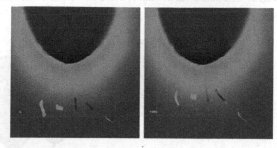

after arriving at the target area, they have to pass cell membrane wall by diffusion process and then the drugs are separated from the drug carrier to cure or kill nuclei of cells. These three steps are required to design size and property of the nanoparticles as a good drug carrier.

However, it is limited to quantify drug carrier efficiency only by experiment. Specifically, in vivo experiment for measuring the mean path of drug carriers is very limited and not sufficient to understand the phenomena of drug carrier transport in human blood flow. At this point, it is helpful to use computational methods for the understanding the behavior of nanoparticles in vascular and extravascular systems. Here, IMFEM is used to calculate the drug carrier path by including red blood cells (RBCs), nanoparticles (NPs) and fluid in a computational domain [4].

Injected NPs undergo complex blood flow by RBC deformation. And then they move to the wall by escaping from the main stream. When they are moving from the center of the blood vessel to the wall, the drug carrier efficiency of NPs can be predicted by quantifying how fast NPs are approaching to the blood vessel wall. It is directly connected to laterally drifting velocity of NPs.

Figure 3 shows the computational domain of blood flow for calculating drift velocity of NPs. A fluid channel is considered with the dimension $20 \times 20 \times 20\,\mu m$. The inlet velocity with $100\,s^{-1}$ linear shear rate is applied at the left end of channel. Each RBC has $7.82\,\mu m$ in diameter and $2.56\,\mu m$ in thickness. Several RBCs are inserted in the fluid domain. And the bottom of the channel is no-slip boundary condition. For more realistic blood flow, the interaction between RBCs is added by Morse potential [7].

In Fig. 4, NPs with 100–1,000 nm diameter are inserted in the fluid domain. As shown in Fig. 4, the linear shear flow in inlet is totally changed by RBC deformation.

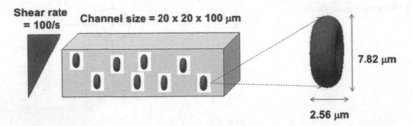

Shear rate = 100/s **Channel size = 20 x 20 x 100 μm**

7.82 μm

2.56 μm

Fig. 3 Computational domain of blood flow for calculating drift velocity of NPs

Fig. 4 Snapshot of NP path under RBC and fluid environment

Furthermore, the complex blood flow makes the NPs drifted to the wall. For the simulation, weak repulsive forces among NPs are required to avoid overlapping between RBC and NP. A Lennard-Jones potential is additionally added to treat molecular interactions.

3 Nanodiamond Platform for Gene Delivery

Researchers have been working on gene delivery to treat diseases associated with defective genes such as cancer, cardio vascular diseases. Since naked DNA (or RNA) is most likely degraded in the blood vessel before it reaches the target cells, successful gene delivery requires the use of a vector that can deliver the DNA into target cells safely as well as efficiently. Nanodiamonds have been investigated intensively because they are promising materials for gene and drug delivery [1, 16]. Recently, it is observed that the transfection efficiency of nanodiamond-based gene delivery is improved 70 times as compared to conventional delivery methods [16]. The above nanodiamond (ND) complex is composed of ND, Polyethylenimine(PEI) 800, and DNA (or RNA). However, the principal mechanism of ND-PEI-DNA interaction is still not clearly understood. In this manuscript, we execute MD simulations in order to investigate interactions of ND-PEI (for functionalized as well as non-functionalized ND).

Fig. 5 Nanodiamond atomic
structure with directionality

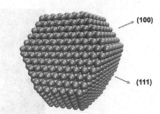

Fig. 6 Atomic structure of
branched PEI 800

Nitrogen : Blue
Carbon : Cyan
Hydrogen : White

Nanodiamonds, which are made of carbon, are manufactured through detonation and several chemical processes. Individual nanodiamonds have a truncated octahedron shape and the diameters ranging from 2 to 6 nm [1]. It has eight hexagonal surfaces with [111] direction and six square surfaces with [100] direction (Fig. 5). Figure 5 illustrates the 4.1 nm ND made of 5,795 atoms. Atoms with [111] direction have −2.23 ec (average) charge while atoms with [100] have 0.13 ec charge (average). The Density Function Tight Binding (DFTB) simulations have been carried out to obtain surface charges of the ND. There are two kinds of PEI's available for the gene delivery applications: Linear PEI and branched PEI. The branched PEI's are used for our MD simulations. The length of the PEI is 3.5 nm, height 1.6 nm and width 1 nm (Fig. 6).

Molecular dynamic simulations are performed to investigate interactions between ND, PEI800. Simulations are carried out with NVT ensemble (300 K) for 1 ns. The functionalized ND has carboxyl group (COOH) on the surface. The carboxyl group produces hydrogen ion (H^+) when mixed with water and results in negative surface charge. The Henderson-Hasselbach relation

$$pK_a = pH + \log\left(\frac{[1-x]}{[x]}\right),$$

is utilized to determine the number of charged site on the ND surface (due to functional group) based on pKa and pH [5]. Here

$$x = \frac{N_{charged}}{N_{total}} = \frac{1}{1 + 10^{pK_a - pH}}, \qquad N_{charged} = \frac{N_{total}}{1 + 10^{pK_a - pH}}$$

x denotes the fraction of ionized site, $N_{charged}$ denotes the number of ionized sites on the ND surface and N_{total} the number of ionizable sites on the ND surface.

Fig. 7 Initial configuration of non-functionalized ND and PEI's (15 Å away from the ND surface)

Fig. 8 Final configuration of non-functionalized ND and PEI's (PEI's do not attach to ND surface)

First, in order to investigate the direct interactions between the non-functionalized ND and PEI, 12 PEI's are placed 15 Å away from the surface of non-functionalized ND (Fig. 7). Since ND-PEI interaction is taking place in water solvent (pH = 7), the simulation box is solvated with water. Since real PEI's are protonated in water, all PEI's in our simulations are 20 % protonated and become cationic (positively charged) polymers [13]. During the equilibration process, the PEI's did not attach to the non-functionalized ND surface since electrostatic interactions between non-functionalized ND and PEI's are too weak to overcome the charge interference of water (Fig. 8).

Next, in order to investigate the interactions between functionalized ND and PEI, 12 PEI's are placed 15 Å away from the surface of functionalized ND (Fig. 9). During the equilibration, the PEI's moves toward ND and attached to the ND surfaces (Fig. 10). These simulation results reveal that functionalized ND can facilitate more PEI's attachments to the ND, which can generate better DNA (RNA) delivery efficiency since the role PEI is to anchor DNA to ND complex.

Fig. 9 Initial configuration of functionalized ND and PEI's (15 Å away from the ND surface)

Fig. 10 Final configuration of functionalized ND and PEI's (PEI's attach to ND surface)

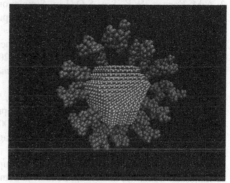

4 Concluding Remarks

Multiscale simulations, ranging from molecular dynamics to finite element method, are performed in order to examine several biomedical applications. First, the IMEFEM framework has been successfully applied to investigate the nanotip and coil electrode sensor system, where it was used to understand the underlying forces present at the nanoscale. Simulations suggest that competing forces due to thermal fluctuations, electric field and molecular interactions play a paramount role in the design of sensory devices. Second, the vascular transport of nanometer scale drug carriers was investigated by using IMEFEM including interactions between RBCs and NPs. From the calculation, it was discovered that the deformations of RBCs in shear flow regime lead to drift NPs to vessel wall. By using this method, drug carrier efficiency during vascular transport can be quantified by simulation. Also, the predicted diffusion coefficients for NPs can be directly used to track NPs in larger vessel like human artery. Lastly, molecular dynamic (MD) simulations are performed in order to investigate the interactions between PEI800 and functionalized nanodiamonds. The MD simulations manifested that functionalized nanodiamond can be a better delivery vector compared to non-functionalized nanodiamond.

Acknowledgements This work was supported by NSF CMMI 0856333 and NSF CMMI 0856492. WKL is supported by World Class University Program through the National Research Foundation of Korea (NRF) funded by the Ministry of Education, Science and Technology (R33-10079).

References

1. A. Adnan, R. Lam, H. Chen, J. Lee, D. Schaffer, A. Barnard, G. Schatz, D. Ho, W.K. Liu, Atomistic simulation and measurement of ph dependent cancer therapeutic interactions with nanodiamond carrier. Mol. Pharm. **8**, 368–374 (2011)
2. M. Gay, L. Zhang, W.K. Liu, Stent modeling using immersed finite element method. Comput. Methods Appl. Mech. Eng. **195**, 4358–4370 (2006)
3. A.M. Kopacz, W.K. Liu, S.Q. Liu, Simulation and prediction of endothelial cell adhesion modulated by molecular engineering. Comput. Methods Appl. Mech. Eng. **197**(25–28), 2340–2352 (2008)
4. A.M. Kopacz, N. Patankar, W.K. Liu, The immersed molecular finite element method. Comput. Methods Appl. Mech. Eng. **233–236**, 28–39 (2012)
5. J. Kotz, P. Treichel, J. Townsend, *Chemistry and Chemical Reactivity* (Brooks Cole, Belmont, 2009)
6. T.R. Lee, Y.S. Chang, J.B. Choi, D.W. Kim, W.K. Liu, Y.J. Kim, Immersed finite element method for rigid body motions in the incompressible navier-stokes flow. Comput. Methods Appl. Mech. Eng. **197**(25–28), 2305–2316 (2008)
7. Y. Liu, W.K. Liu, Rheology of red blood cell aggregation by computer simulation. J. Comput. Phys. **220**(1), 139–154 (2006)
8. Y. Liu, L. Zhang, X. Wang, W.K. Liu, Coupling of navier-stokes equations with protein molecular dynamics and its application to hemodynamics. Int. J. Numer. Methods Fluids **46**, 1237–1252 (2004)
9. Y. Liu, J.H. Chung, W.K. Liu, R. Ruoff, Dielectrophoretic assembly of nanowires. J. Phys. Chem. B **110**(29), 14098–14106 (2006)
10. W.K. Liu, D.W. Kim, S. Tang, Mathematical foundations of the immersed finite element method. Comput. Mech. **39**, 211–222 (2006)
11. W.K. Liu, Y. Liu, D. Farrell, L. Zhang, X.S. Wang, Y. Fukui, N. Patankar, Y. Zhang, C. Bajaj, J. Lee, J. Hong, X. Chen, H. Hsua, Immersed finite element method and its applications to biological systems. Comput. Methods Appl. Mech. Eng. **195**, 1722–1749 (2006)
12. Y. Liu, W.K. Liu, T. Belytschko, N. Patankar, A.C. To, A.M. Kopacz, J.H. Chung, Immersed electrokinetic finite element method. Int. J. Numer. Methods Eng. **71**, 379–405 (2007)
13. P. Uthe, The development of polycationic materiala for gene delivery applications, Ph.D. dissertation, University of North Carolina, Chapel Hill (2010)
14. X. Wang, W.K. Liu, Extended immersed boundary method using fem and rkpm. Comput. Methods Appl. Mech. Eng. **193**(12–14), 1305–1321 (2004)
15. L. Zhang, A. Gerstenberger, X. Wang, W.K. Liu, Immersed finite element method. Comput. Methods Appl. Mech. Eng. **193**(21–22), 2051–2067 (2004)
16. X.Q. Zhang, M. Chen, R. Lam, X. Xu, E. Osawa, D. Ho, Polymer-functionalized nanodiamond platforms as vehicles for gene delivery. ACS Nano **3**, 2609–2616 (2009)

Corrected Stabilized Non-conforming Nodal Integration in Meshfree Methods

Marcus Rüter, Michael Hillman, and Jiun-Shyan Chen

Abstract A novel approach is presented to correct the error from numerical integration in Galerkin methods for meeting linear exactness. This approach is based on a Ritz projection of the integration error that allows a modified Galerkin discretization of the original weak form to be established in terms of assumed strains. The solution obtained by this method is the correction of the original Galerkin discretization obtained by the inaccurate numerical integration scheme. The proposed method is applied to elastic problems solved by the reproducing kernel particle method (RKPM) with first-order correction of numerical integration. In particular, stabilized non-conforming nodal integration (SNNI) is corrected using modified ansatz functions that fulfill the linear integration constraint and therefore conforming sub-domains are not needed for linear exactness. Illustrative numerical examples are also presented.

Keywords Reproducing kernel particle method • Stabilized non-conforming nodal integration • Integration constraint • Strain smoothing

1 Introduction

As the name implies, meshfree methods are based on a discretization of the continuous problem without using a mesh, as opposed to the finite element method and related mesh-based methods. Meshfree methods therefore have several obvious advantages, especially when the mesh is the source of problems, e.g. in the large deformation analysis of structures or in the simulation of discontinuities such as cracks.

M. Rüter (✉) · M. Hillman · J.-S. Chen
Department of Civil and Environmental Engineering, University of California,
Los Angeles, CA, 90095, USA
e-mail: mruter@seas.ucla.edu; hillman@seas.ucla.edu; jschen@seas.ucla.edu

M. Griebel and M.A. Schweitzer (eds.), *Meshfree Methods for Partial Differential Equations VI*, Lecture Notes in Computational Science and Engineering 89, DOI 10.1007/978-3-642-32979-1_5, © Springer-Verlag Berlin Heidelberg 2013

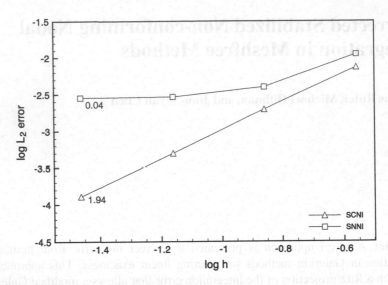

Fig. 1 Failure of SNNI for the tube problem discussed in Sect. 5.2

A crucial issue in Galerkin meshfree methods is domain integration, typically carried out by Gauß or nodal quadrature. In the former case, integration is performed using a regular cell structure or background mesh. Here, significant integration errors arise when the background mesh is not aligned with shape function supports, see [10]. Nodal integration, on the other hand, yields low convergence rates since the integrands are poorly integrated, and it also suffers from rank instability due to the shape function derivatives nearly vanishing at the particles. To cope with these problems Beissel and Belytschko [4] suggested a stabilization method for nodal integration that relies on a least squares residual term which is added to the total potential energy functional. Bonet and Kulasegaram [5] added terms to the derivatives of the shape functions to obtain an improved method. Chen et al. [8, 9, 13] proposed stabilized conforming nodal integration (SCNI), based on gradient smoothing which passes the linear patch test. However, its non-conforming counterpart termed stabilized non-conforming nodal integration (SNNI), see [7], is often used in practice because of its simplicity, particularly for fragment-impact type problems where the construction of conforming strain smoothing sub-domains is tedious, see [11]. The use of non-conforming strain smoothing with divergence operation in SNNI fails to pass the linear patch test, and integration error becomes an issue. Figure 1 illustrates the failure of SNNI. In this figure, the convergence curves for SCNI and SNNI are compared for the tube inflation problem, as presented in Sect. 5.2, and the low convergence rate in the case of SNNI is apparent.

Whenever a Galerkin discretization is used to find an approximate solution of the model problem at hand, its discretization error (the difference between the true solution and its Galerkin approximation) has to be controlled to obtain reliable numerical results. More precisely, error control means that the error is estimated by a reliable a posteriori error estimator. This is in general the best that can be done, since the error itself is in the infinite-dimensional trial and test space and therefore not easy to grasp. For further details on discretization error control the reader may consult Stein and Rüter [14] and Babuška et al. [3]. In addition to the discretization error, for numerical integration schemes such as SNNI the integration error (the difference between the true Galerkin solution and its quadrature approximation) has to be taken into account. However, since the integration error is in the same finite-dimensional trial and test space as both the true Galerkin solution and its quadrature approximation, there is, in principle, a chance to compute it exactly. Therefore, the integration error can be corrected rather than just estimated, which is demonstrated in this paper for the case of SNNI.

Babuška et al. [1] propose an energy norm a priori error estimate for the combined discretization and integration error (the difference between the true solution and its Galerkin quadrature approximation), provided that the integration error was corrected by a simple correction of the stiffness matrix to fulfill the so-called zero row-sum condition. This estimate was later extended by Babuška et al. [2] from linear basis functions to higher-order basis functions.

In this paper, we devise a scheme to correct the integration error induced from using SNNI. This correction approach is based on a modification of the test functions to fulfill the linear integration constraint as required to pass the linear patch test. With the corrected test functions, it becomes possible, under certain assumptions, to compute the true Galerkin solution using a modified bilinear form with SNNI.

The paper is organized as follows: in Sect. 2, the model problem of linear elasticity in its strong and weak forms is presented, as well as the associated Galerkin discretization. In Sect. 3, the correction approach to correct the error from numerical integration is introduced using the true Galerkin solution as the Ritz projection of the integration error. The correction is then applied to meshfree methods in Sect. 4, specifically to RKPM with SNNI employed. Numerical examples are presented in Sect. 5, where the correction approach is applied to problems in linear elasticity. The paper concludes with Sect. 6 which summarizes the major findings achieved from theoretical and numerical points of view.

2 The Model Problem of Linear Elasticity

This section gives a brief account of the model problem of linear elasticity in its strong and weak forms as well as its Galerkin discretization in an abstract setting.

2.1 Strong Form of the Model Problem

Let an isotropic, linear elastic body be given by the open, bounded, polygonal or polyhedral domain $\Omega \subset \mathbb{R}^d$ with dimension d. Its boundary $\Gamma = \partial\Omega$ consists of two disjoint parts such that $\Gamma = \bar{\Gamma}_D \cup \bar{\Gamma}_N$ with Γ_D and Γ_N being the portions of Γ where Dirichlet and Neumann boundary conditions are imposed, respectively.

The strong form of the elliptic, self-adjoint model problem of linear elasticity is to find the displacement field u such that the field equations

$$-\operatorname{div}\sigma(u) = f \quad \text{in } \Omega \tag{1a}$$

$$\sigma - \mathbb{C} : \varepsilon(u) = 0 \quad \text{in } \Omega \tag{1b}$$

$$\varepsilon - \nabla^{\text{sym}} u = 0 \quad \text{in } \Omega \tag{1c}$$

subjected to the boundary conditions

$$u = \bar{u} \quad \text{on } \Gamma_D \tag{1d}$$

$$\sigma(u) \cdot n = \bar{t} \quad \text{on } \Gamma_N \tag{1e}$$

are fulfilled. Here, the given body forces f and the prescribed boundary tractions $\bar{t}$ are assumed to be in $L_2(\Omega)$ and $L_2(\Gamma_N)$, respectively. Furthermore, σ is the stress tensor, and ε is the strain tensor, which are related via the elasticity tensor $\mathbb{C}$ that depends on Young's modulus E and Poisson's ratio ν.

2.2 Weak Form of the Model Problem

In the weak form associated with (1a) we seek $u \in \mathcal{V} = \{v \in H^1(\Omega) ; v_{|\Gamma_D} = \bar{u}\}$ such that

$$a(u, v) = F(v) \quad \forall v \in \mathcal{V}_0 \tag{2}$$

with $\mathcal{V}_0 = \{v \in H^1(\Omega) ; v_{|\Gamma_D} = 0\}$. Moreover, $a : \mathcal{V} \times \mathcal{V}_0 \to \mathbb{R}$ and $F : \mathcal{V}_0 \to \mathbb{R}$ are bilinear and linear forms defined as

$$a(u, v) = \int_\Omega \sigma(u) : \varepsilon(v) \, dV \tag{3}$$

and

$$F(v) = \int_\Omega f \cdot v \, dV + \int_{\Gamma_N} \bar{t} \cdot v \, dA, \tag{4}$$

respectively. Since a is coercive, and both a and F are bounded, $u \in \mathcal{V}$ exists and is the unique solution to (2) owing to the Lax-Milgram theorem.

2.3 Galerkin Discretization

In order to solve the weak form (2) using a Galerkin method, e.g. either a mesh-based method such as the finite element method (FEM) or a meshfree method such

as RKPM, the Galerkin discretization of (2) is introduced. For this, the variational
problem (2) is projected onto a finite-dimensional subspace $\mathcal{V}_h$ of $\mathcal{V}$ and we solve

$$a(u_h, v_h) = F(v_h) \quad \forall v_h \in \mathcal{V}_{0,h} \subset \mathcal{V}_0 \tag{5}$$

for an approximate solution $u_h \in \mathcal{V}_h$. Since $\mathcal{V}_h \subset \mathcal{V}$, the approximate solution u_h
exists in $\mathcal{V}_h$ and is unique. Another conclusion that can be drawn from $\mathcal{V}_{0,h} \subset \mathcal{V}_0$ is
that $v_h \in \mathcal{V}_0$ and thus we may subtract (5) from (2) to see that

$$a(u - u_h, v_h) = 0 \quad \forall v_h \in \mathcal{V}_{0,h}, \tag{6}$$

which is the well-known Galerkin orthogonality condition, meaning that the
discretization error $u - u_h$ is orthogonal to the test space $\mathcal{V}_{0,h}$.

3   Correction of the Numerical Integration Error

In this section, numerical integration in Galerkin methods is discussed, and a method
to provide the correction (with specific form for the first-order correction shown in
Sect. 4) of the associated integration error is derived to recover the true Galerkin
solution that possesses the important Galerkin orthogonality relation.

3.1   Numerical Integration in Galerkin Methods

Whenever a Galerkin method is employed, numerical integration may become
an issue. This is especially true for meshfree methods, owing to the overlapping
supports of shape functions and the rational functions that need to be integrated.
This is also true for mesh-based Galerkin methods, such as the extended finite
element method (XFEM), where the derivatives of the enrichment functions have
singularities at the crack tip. In these cases, we can only approximate a and F. We
denote these approximations by a_h and F_h, respectively. Substituting a_h and F_h into
(5), we now search for a solution u_{hh} as an approximation of u_h in the same space
$\mathcal{V}_h$ such that

$$a_h(u_{hh}, v_h) = F_h(v_h) \quad \forall v_h \in \mathcal{V}_{0,h}. \tag{7}$$

If the coercivity of a_h, and boundedness of a_h and F_h are not lost by the
numerical integration, u_{hh} is the unique solution to (7) in $\mathcal{V}_h$. However, the Galerkin
orthogonality (6) is violated in this case, i.e. $a(u - u_{hh}, v_h) \neq 0$ for all $v_h \in \mathcal{V}_{0,h}$,
since (7) is not a Galerkin discretization of (2). The Galerkin discretization (5) of
the weak form (2) was obtained by replacing the trial and test spaces $\mathcal{V}$ and $\mathcal{V}_0$ with
$\mathcal{V}_h$ and $\mathcal{V}_{0,h}$, respectively, while keeping the same model in the sense of a and F.
Conversely, (7) is obtained from the Galerkin discretization (5) by replacing a and
F with a_h and F_h, respectively, while keeping the same spaces $\mathcal{V}_h$ and $\mathcal{V}_{0,h}$.

3.2 Correction of the Integration Error

In order to recover the true Galerkin solution u_h, the general idea in SCNI, see [8], is to modify the bilinear form a_h in such a way that u_h becomes the solution of (7) rather than its approximation u_{hh}. A similar idea is employed herein.

Let us first add and subtract u_h in a_h to see that (7) can be recast into

$$a_h(u_h, v_h) - a_h(u_h - u_{hh}, v_h) = F_h(v_h) \quad \forall v_h \in \mathcal{V}_{0,h}. \tag{8}$$

The above variational problem can be interpreted as follows: in order to recover u_h from (7), we propose to correct $a_h(u_h, v_h)$ by the integration error represented as $a_h(u_h - u_{hh}, v_h)$. However, to obtain the exact Galerkin solution from (8), it is necessary that the first argument in the second term of the left-hand side also depends on u_h only. To this end, let us introduce the bilinear form $\overset{\Delta}{a}_h : \mathcal{V}_h \times \mathcal{V}_{0,h} \to \mathbb{R}$ defined as

$$\overset{\Delta}{a}_h(u_h, v_h) = -a_h(u_h - u_{hh}, v_h) \quad \forall v_h \in \mathcal{V}_{0,h}, \tag{9}$$

i.e. we introduce u_h as the Ritz projection of $u_h - u_{hh}$. The bilinear form $\overset{\Delta}{a}_h$ can be designed such that the first argument is the same as in a_h, whereas the integration error is embedded in the second argument and thus $\overset{\Delta}{a}_h$ takes the general form

$$\overset{\Delta}{a}_h(u_h, v_h) = \fint_\Omega \sigma(u_h) : \overset{\Delta}{\hat{\varepsilon}}(v_h)\, dV. \tag{10}$$

Here, the bar in the integral sign represents numerical integration, and $\overset{\Delta}{\hat{\varepsilon}}$ is an assumed strain tensor that takes the effect of the integration error into account as we shall see later in Sect. 4.3. With the definition (9), the weak form (8) turns into

$$a_h(u_h, v_h) + \overset{\Delta}{a}_h(u_h, v_h) = F_h(v_h) \quad \forall v_h \in \mathcal{V}_{0,h}, \tag{11}$$

or, more concisely, into

$$\hat{a}_h(u_h, v_h) = F_h(v_h) \quad \forall v_h \in \mathcal{V}_{0,h}, \tag{12}$$

which we can now solve for $u_h \in \mathcal{V}_h$ directly. In the above, the bilinear form $\hat{a}_h : \mathcal{V}_h \times \mathcal{V}_{0,h} \to \mathbb{R}$ is defined as $\hat{a}_h(\cdot, \cdot) = a_h(\cdot, \cdot) + \overset{\Delta}{a}_h(\cdot, \cdot)$, which, thanks to (3) and (10), can be expressed as

$$\hat{a}_h(u_h, v_h) = \fint_\Omega \sigma(u_h) : \hat{\varepsilon}(v_h)\, dV \tag{13}$$

in terms of the assumed strain tensor $\hat{\varepsilon} = \varepsilon + \overset{\Delta}{\hat{\varepsilon}}$. At first sight, the correction seems to be computationally expensive, since it appears to involve solving two global systems. However, (12) is the only problem we need to solve, since it replaces the original problem (7). Furthermore, the stiffness matrix associated with the bilinear form $\hat{a}_h$ requires approximately the same computational costs as a_h as will be shown later in Sect. 4.3.

Note that since $\hat{a}_h$ differs from a_h by the second argument only, (12) can be interpreted as a Petrov-Galerkin correction of (7). Therefore, the associated stiffness matrix is unsymmetric and thus requires a different class of iterative solvers. Nevertheless, if a_h is coercive and bounded, so is $\hat{a}_h$ provided that (9) holds. Consequently, the solution u_h to the problem (12) exists and is unique. Moreover, since u_h also satisfies (5), the Galerkin orthogonality (6) is recovered. As mentioned above, in other correction methods such as SCNI the stiffness matrix is symmetric, since in SCNI assumed strains are used for both arguments in the bilinear form $\hat{a}_h$, see [13].

If the integration method is accurate and $u_{hh} = u_h$, i.e. the integration error vanishes, the method is consistent in the sense that nothing has to be corrected and it turns out that $\hat{a}_h = a_h = a$ and thus the associated stiffness matrix becomes symmetric.

It should finally be noted that the recovered Galerkin solution u_h and the recovered Galerkin orthogonality (6) allow the estimation of the discretization error with the various error estimation techniques developed over the years. If the integration error could not be corrected, estimation of the discretization error would be much more cumbersome, since the Galerkin orthogonality is an important requirement used in most a posteriori discretization error estimates.

4   Meshfree Methods

All that remains to show is how the assumed strain tensor $\hat{\varepsilon}$ can be constructed for a specific Galerkin method. In this paper, we consider the reproducing kernel particle method as representative of meshfree Galerkin methods. Emphasis is placed on its numerical integration using stabilized non-conforming nodal integration.

4.1   The Reproducing Kernel Particle Method (RKPM)

In RKPM, the approximate solution u_h can be expressed in terms of the following form

$$u_h = \sum_{n_P} \Psi_I(x) u_I, \tag{14}$$

where n_P is the number of particles x_I, i.e. $n_P = \text{card}\{x_I ; x_I \in \bar{\Omega}\}$, u_I are the coefficients, and Ψ_I are the meshfree shape functions. In the case of the reproducing kernel particle method, Ψ_I takes the form

$$\Psi_I(x) = \Phi(x - x_I) H^T(0) M^{-1}(x) H(x - x_I) \tag{15}$$

with the kernel function Φ, the vector of monomial basis functions H, and the moment matrix

$$M(x) = \sum_{nP} \Phi(x - x_I) H(x - x_I) H^T(x - x_I). \qquad (16)$$

In this paper, linear basis functions are employed facilitating linear completeness which is one of the requirements to achieve linear exactness in the Galerkin approximation.

Since, in general, meshfree Galerkin approximations are not kinematically admissible, the essential boundary conditions need to be imposed carefully, e.g. by the Lagrange multiplier method, the penalty method or Nitsche's method. Here, we review the latter as introduced in [12] to impose the Dirichlet boundary conditions (1d) in a weak sense. In this case, the bilinear form a and the linear form F, as defined in (3) and (4), are extended to

$$\begin{aligned} a(u, v) = \int_{\Omega} \sigma(u) : \varepsilon(v) \, dV - \int_{\Gamma_D} v \cdot \sigma(u) \cdot n \, dA \\ - \int_{\Gamma_D} u \cdot \sigma(v) \cdot n \, dA + \int_{\Gamma_D} \beta u \cdot v \, dA \end{aligned} \qquad (17)$$

and

$$F(v) = \int_{\Omega} f \cdot v \, dV + \int_{\Gamma_N} \bar{t} \cdot v \, dA - \int_{\Gamma_D} \bar{u} \cdot \sigma(v) \cdot n \, dA + \int_{\Gamma_D} \beta \bar{u} \cdot v \, dA, \qquad (18)$$

respectively, with parameter $\beta \in \mathbb{R}^+$ to ensure coercivity of a. As a consequence of the weak fulfillment of the Dirichlet boundary conditions, the trial and test space now takes the simple form $\mathcal{V} = H^1(\Omega)$. With the RKPM approximation (14) at hand, we may now define the finite-dimensional subspace $\mathcal{V}_h \subset \mathcal{V}$ as $\mathcal{V}_h = \{v_h \in H^1(\Omega) ; v_h \text{ as in (14)}\}$.

4.2 Stabilized Conforming and Non-conforming Nodal Integration

A crucial point in numerical integration is the fulfillment of the associated linear integration constraint obtained by satisfying (7) with linear solution, see [7] for details,

$$\oint_{\Omega} \nabla \Psi_I \, dV = \oint_{\partial\Omega} \Psi_I n \, dA \qquad (19)$$

which is necessary to achieve linear exactness in the Galerkin approximation and therefore to pass the linear patch test. Linear exactness can also be interpreted as the linear portion of the integration error vanishing. We remark that for high-order exactness, high-order integration constraints need to be met. For further details on the construction of generalized integration constraints we refer the reader to Chen et al. [6].

To keep the meshfree nature of RKPM, in this paper nodal integration is considered rather than Gauß integration to evaluate the integrals in (17) and (18). In nodal integration, the integrals are evaluated at the particles and weighted with a representative integration domain. It is well known, however, that nodal integration yields rank instability and low convergence rates.

A remedy for both problems is obtained using smoothed gradients for nodal integration introduced by Chen et al. [8], rather than standard gradients. The smoothed gradient operator $\tilde{\nabla}$ is defined as

$$\tilde{\nabla}(\cdot)|_{x_L} = \frac{1}{|\Omega_L|} \int_{\Omega_L} \nabla(\cdot)|_x \, dV = \frac{1}{|\Omega_L|} \int_{\partial\Omega_L} (\cdot)|_x \, \boldsymbol{n} \, dA, \tag{20}$$

where Ω_L is the representative domain of node L, which is also referred to as smoothing domain. Here, the gradient is first smoothed over the domain Ω_L, and the divergence theorem is then applied to convert the domain integral to a surface integral.

In the case that the representative domains Ω_L are conforming, i.e. they are disjoint and $\bar{\Omega} = \bigcup_{np} \bar{\Omega}_L$, e.g. using Voronoi cells, it is easily verified using definition (20) and nodal integration with weight $|\Omega_L|$ that

$$\oint_{\Omega} \tilde{\nabla}\Psi_I \, dV = \sum_{np} \tilde{\nabla}\Psi_I(\boldsymbol{x}_L)|\Omega_L| = \sum_{np} \oint_{\partial\Omega_L} \Psi_I \boldsymbol{n} \, dA$$
$$= \oint_{\partial\Omega} \Psi_I \boldsymbol{n} \, dA \tag{21}$$

holds and thus the linear integration constraint (19) is fulfilled for the smoothed gradient operator $\tilde{\nabla}$. Owing to the conforming nature of the integration domains Ω_L as depicted in Fig. 2, this method is termed stabilized conforming nodal integration. This integration method, however, has the drawback that the conforming domains are sometimes difficult to construct, e.g. in a semi-Lagrangian formulation for fragment-impact problems, see [11]. To cope with this problem, stabilized non-conforming nodal integration has been introduced in [7, 11]. As the name implies, the domains Ω_L are not conforming as shown in Fig. 3, which is a simplicity taken in order to overcome the difficulty of constructing conforming domains. For this method, the last equality in (21) no longer holds since the domains are not conforming, and as a consequence the linear integration constraint (19) is not fulfilled which may result in considerable integration error and deterioration of convergence rate.

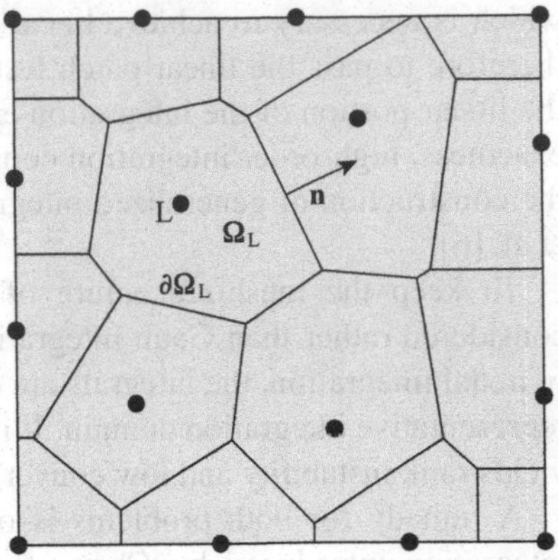

Fig. 2 Smoothing domains Ω_L in SCNI using Voronoi cells

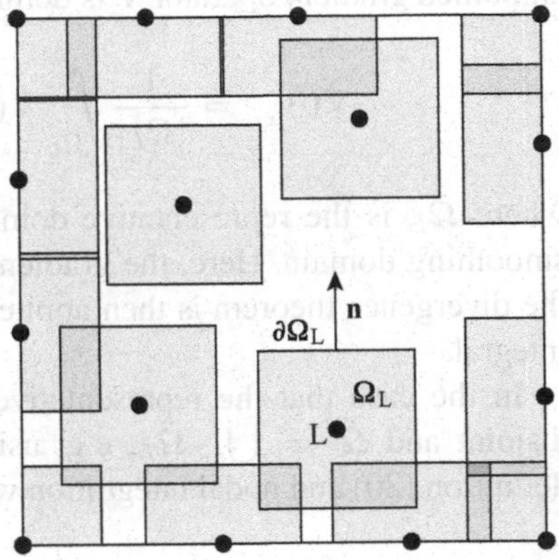

Fig. 3 Smoothing domains Ω_L in SNNI

4.3 The Assumed Strain Tensor $\hat{\varepsilon}$

As we have seen, the linear integration constraint (19) is fulfilled by SCNI. We therefore examine the first-order correction of SNNI to pass the linear patch test. To satisfy the linear integration constraint (19) for SNNI, we introduce the modified ansatz function (for the gradients)

$$\nabla \hat{\Psi}_I = \nabla \Psi_I + \underbrace{\sum_d \xi_{dI} \nabla \hat{\Psi}_{dI}}_{=\nabla \overset{\triangle}{\Psi}_I} \tag{22}$$

in terms of gradients of the standard ansatz function $\nabla \Psi_I$ and a first-order correction term $\nabla \overset{\triangle}{\Psi}_I$ to account for the integration error which depends on the unknown coefficients ξ_{dI} and the gradient of the correction ansatz functions $\nabla \hat{\Psi}_{dI}$. It should be clear that since the test space $\mathscr{V}_h$ is constructed in terms of the modified ansatz functions $\hat{\Psi}_I$, the choice for the first-order correction ansatz functions $\hat{\Psi}_{dI}$ is restricted.

As an example, we express $\nabla \hat{\Psi}_{dI}$ in terms of the shape function supports

$$\Theta_I(x) = \begin{cases} 1 & \text{if } x \in \text{supp } \Psi_I(x) \\ 0 & \text{if } x \notin \text{supp } \Psi_I(x) \end{cases}. \tag{23}$$

In order to determine $\nabla \overset{\Delta}{\Psi}_I$ and thus to be able to compute $\overset{\Delta}{\varepsilon}$ and consequently the assumed strain tensor $\hat{\varepsilon}$, we substitute (22) into the linear integration constraint (19) to see that

$$\fint_\Omega \left\{ \begin{pmatrix} \Psi_{I,x} \\ \Psi_{I,y} \end{pmatrix} + \xi_{1I} \begin{pmatrix} \Theta_I \\ 0 \end{pmatrix} + \xi_{2I} \begin{pmatrix} 0 \\ \Theta_I \end{pmatrix} \right\} dV = \fint_{\partial\Omega} \begin{pmatrix} \Psi_I n_1 \\ \Psi_I n_2 \end{pmatrix} dA \tag{24}$$

holds with additional terms arising from the modified gradient in (22) and the shape function supports (23). In the above, $\Psi_{I,x}$ and $\Psi_{I,y}$ are the derivatives of Ψ_I with respect to x and y, respectively. The unknown coefficients ξ_{1I} and ξ_{2I} can then be determined as

$$\xi_{1I} = \left\{ \fint_\Omega \Theta_I \, dV \right\}^{-1} \left\{ \fint_{\partial\Omega} \Psi_I n_1 \, dA - \fint_\Omega \Psi_{I,x} \, dV \right\} \tag{25a}$$

$$\xi_{2I} = \left\{ \fint_\Omega \Theta_I \, dV \right\}^{-1} \left\{ \fint_{\partial\Omega} \Psi_I n_2 \, dA - \fint_\Omega \Psi_{I,y} \, dV \right\}. \tag{25b}$$

Note that the computation of the coefficients ξ_{1I} and ξ_{2I} is inexpensive, since it can be done locally. This is in contrast to previous integration corrections presented in [5, 13], where a global system needs to be solved. Furthermore, the proposed algorithm to compute ξ_{1I} and ξ_{2I} can be easily implemented into an existing RKPM code.

We are now in a position to define the assumed strain tensor $\hat{\varepsilon}$ used in the bilinear form $\overset{\Delta}{a}_h$ in (13). With (22) the tensor takes the form

$$\hat{\varepsilon}(v_h) = \sum_{nP} \nabla^{\text{sym}}[\hat{\Psi}_I(x)v_I]. \tag{26}$$

Note that for imposing Dirichlet boundary conditions in a weak sense using Nitsche's method, we further need to compute assumed stresses $\hat{\sigma}(v_h)$ in (17) and (18), which can be done by the constitutive relation $\hat{\sigma}(v_h) = \mathbb{C} : \hat{\varepsilon}(v_h)$.

With the assumed strain tensor $\hat{\varepsilon}$ at hand, the bilinear form $\hat{a}_h$ as defined in (13) can be computed. Thus, the modified variational problem (12) can be solved for the Galerkin solution u_h to pass the linear patch test. From the construction of $\hat{\varepsilon}$ in (26) it is clear that when the linear integration constraint (19) is met, a problem with linear solution can be recovered. For the recovery of high-order solutions and for use with different integration methods, such as Gauß integration or direct nodal integration (DNI), we refer to the unified framework for domain integration as recently presented by Chen et al. [6].

5 Numerical Examples

In this section, we present numerical results obtained by the first-order correction of integration presented in the preceding sections. In particular, we focus on SNNI and its first-order correction and compare the results with uncorrected SNNI and SCNI.

5.1 Beam Problem

In our first example, the system is a plane-stress cantilever beam subjected to a parabolic in-plane shear traction with the total load $P = 2,000$ kN on the free end, as depicted in Fig. 4. The beam is modeled using anti-symmetric boundary conditions on bottom of the beam. The length of the beam is $L = 10$ m, its height is $H = 2$ m and it is made of an isotropic, linear elastic material with the properties: Young's modulus $E = 30 \cdot 10^6$ kN/m^2 and Poisson's ratio $\nu = 0.3$.

The sub-domains of the strain smoothing used for SCNI and SNNI are plotted in Fig. 5. As can be observed, the smoothing zones are much easier to construct for SNNI, since SCNI requires Voronoi cells.

The deflection error along the centroid of the beam is shown in Fig. 6. First-order correction of SNNI shows similar performance as SCNI with tip displacement accuracies shown in Table 1.

Next, convergence of the methods is considered with the uniform node refinements of the half beam as shown in Fig. 7 using 51, 165, 585, and 2,193 nodes. The model parameters for this example are $L = 5$ m and $H = 2$ m.

The associated convergence is plotted in Fig. 8. As can be seen, corrected SNNI shows a large improvement in error as well as convergence rate. It should be noted, however, that the true Galerkin solution cannot be fully recovered in this case, since the proposed correction restores linear solutions only and thus there is still integration error left.

Finally, we increase Poisson's ratio to $\nu = 0.49999999$ to let the material be nearly incompressible. Furthermore, we let $L = 10$ m, $H = 2$ m, and use a plane-strain structure. Even in this case corrected SNNI yields very good results with nearly vanishing deflection error as can be observed from Fig. 9. The reason is that nodal integration acts, by construction, as an under integration method and thus locking does not become an issue, whereas Gauß integration with 5×5 integration points yields locking.

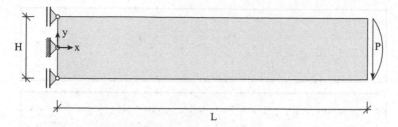

Fig. 4 System and loading for the cantilever beam problem

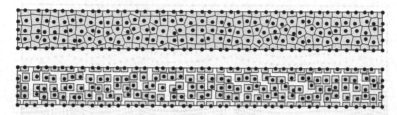

Fig. 5 Smoothing schemes for SCNI (*top*) and SNNI (*bottom*)

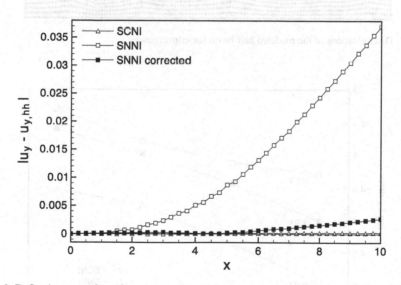

Fig. 6 Deflection error along the centroid of the beam

Table 1 Tip displacement accuracy (numerical solution normalized by the analytical solution)

Method	Uncorrected	Corrected
SCNI	100.62 %	–
SNNI	110.75 %	101.75 %

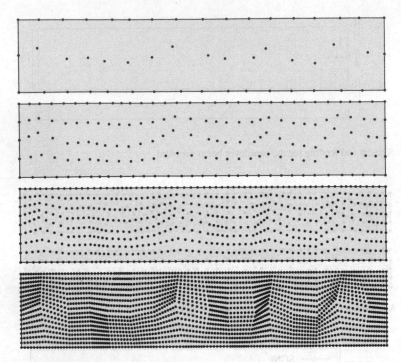

Fig. 7 Discretizations of the modeled half beam for convergence

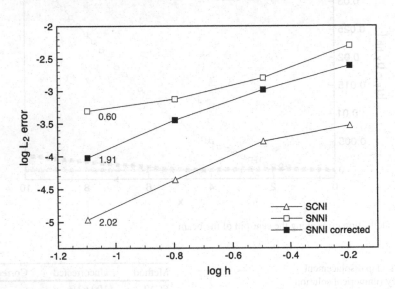

Fig. 8 Convergence for the beam problem

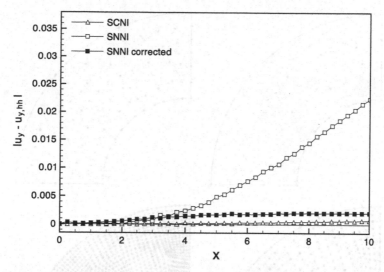

Fig. 9 Deflection error along the centroid of a nearly incompressible beam

Fig. 10 A plane-strain tube subjected to an internal pressure

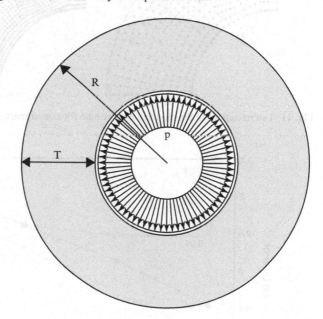

5.2 Tube Problem

In the next example, the structure is an isotropic tube in a plane-strain state with outer radius $R = 1\,\text{m}$ and thickness $T = 0.5\,\text{m}$ subjected to internal pressure $p = 10 \cdot 10^6\,\text{kN/m}^2$, as shown in Fig. 10. Due to symmetry conditions, only one quarter of the domain is modeled with anti-symmetric boundary conditions. The material data is the same as in the previous problem for the compressible case.

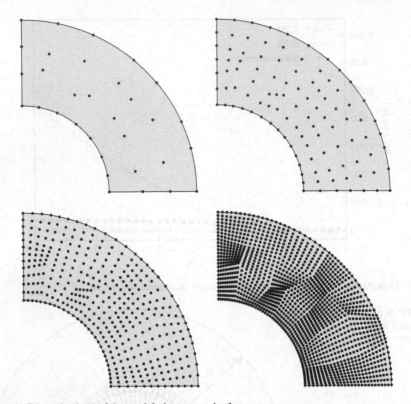

Fig. 11 Discretizations of the modeled quarter tube for convergence

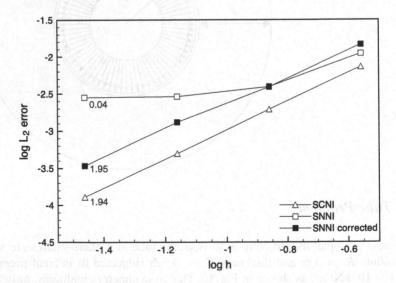

Fig. 12 Convergence for the tube problem

The modeled domain is discretized by 32, 105, 377, and 1,425 nodes with irregular node distributions, as depicted in Fig. 11.

The convergence curves in this example, as plotted in Fig. 12, show similar behavior as in the previous example, i.e. without first-order correction SNNI has a low convergence rate, whereas with the correction the convergence rate is restored and is similar to the result from SCNI.

6 Conclusions

A method was derived to correct the integration error in Galerkin methods and therefore recover the true Galerkin approximation of the continuous model problem at hand. The cornerstone of the proposed correction method is to modify the bilinear form of the original method using the same numerical integration scheme. This modified bilinear form can be viewed as accounting for the integration error by using test functions that fulfill the associated requirements for exactness. In particular, the method was applied to recover linear exactness under the framework of reproducing kernel approximations with stabilized non-conforming nodal integration, which is known to fail the linear integration constraint. Further, the proposed assumed strain to meet integration constraint only requires solving a small linear system locally. Numerical examples showed that the correction method performs well even for nearly incompressible elastic problems, and thus offers a viable alternative to constructing conforming cells previously used in SCNI.

Acknowledgements The support of this work by the US Army Engineer Research and Development Center under the contract W912HZ-07-C-0019:P00001 to the second and third authors and DFG (German Research Foundation) under the grant no. RU 1213/2-1 to the first author is very much appreciated.

References

1. I. Babuška, U. Banerjee, J.E. Osborn, Q. Li, Quadrature for meshless methods. Int. J. Numer. Methods Eng. **76**, 1434–1470 (2008)
2. I. Babuška, U. Banerjee, J.E. Osborn, Q. Zhang, Effect of numerical integration on meshless methods. Comput. Methods Appl. Mech. Eng. **198**, 2886–2897 (2009)
3. I. Babuška, J. Whiteman, T. Strouboulis, *Finite Elements: An Introduction to the Method and Error Estimation* (Oxford University Press, Oxford, 2010)
4. S. Beissel, T. Belytschko, Nodal integration of the element-free Galerkin method. Comput. Methods Appl. Mech. Eng. **139**, 49–74 (1996)
5. J. Bonet, S. Kulasegaram, Correction and stabilization of smooth particle hydrodynamics methods with applications in metal forming simulations. Int. J. Numer. Methods Eng. **47**, 1189–1214 (2000)
6. J.S. Chen, M. Hillman, M. Rüter, *A unified domain integration method for Galerkin meshfree methods*, submitted to Int. J. Numer. Methods. Eng., (2012)

7. J.S. Chen, W. Hu, M.A. Puso, Y. Wu, X. Zhang, Strain smoothing for stabilization and regularization of Galerkin meshfree methods, in *Meshfree Methods for Partial Differential Equations III*, ed. by M. Griebel, M.A. Schweitzer (Springer, Berlin, 2007), pp. 57–75

8. J.S. Chen, C.-T. Wu, S. Yoon, Y. You, A stabilized conforming nodal integration for Galerkin mesh-free methods. Int. J. Numer. Methods Eng. **50**, 435–466 (2001)

9. J.S. Chen, S. Yoon, C.-T. Wu, Non-linear version of stabilized conforming nodal integration for Galerkin mesh-free methods. Int. J. Numer. Methods Eng. **53**, 2587–2615 (2002)

10. J. Dolbow, T. Belytschko, Numerical integration of the Galerkin weak form in meshfree methods. Comput. Mech. **23**, 219–230 (1999)

11. P.C. Guan, S.W. Chi, J.S. Chen, T.R. Slawson, M.J. Roth, Semi-Lagrangian reproducing kernel particle method for fragment-impact problems. Int. J. Impact Eng. **38**, 1033–1047 (2011)

12. J. Nitsche, Über ein Variationsprinzip zur Lösung von Dirichlet-Problemen bei Verwendung von Teilräumen, die keinen Randbedingungen unterworfen sind. Abh. Math. Semin. Univ. Hambg. **36**, 9–15 (1971)

13. M.A. Puso, J.S. Chen, E. Zywicz, W. Elmer, Meshfree and finite element nodal integration methods. Int. J. Numer. Methods Eng. **74**, 416–446 (2008)

14. E. Stein, M. Rüter, Finite element methods for elasticity with error-controlled discretization and model adaptivity, in *Encyclopedia of Computational Mechanics*, 2nd edn., ed. by E. Stein, R. de Borst, T.J.R. Hughes (Wiley, Chichester, 2007)

Multilevel Partition of Unity Method for Elliptic Problems with Strongly Discontinuous Coefficients

Marc Alexander Schweitzer

Abstract In this paper, we study the robustness of a multilevel partition of unity method. To this end, we consider a scalar diffusion equation in two and three space dimensions with large jumps in the diffusion coefficient or material properties. Our main focus in this investigation is if the use of simple enrichment functions is sufficient to attain a robust solver independent of the geometric complexity of the material interface.

Keywords Multilevel methods • Robustness • Jumping coefficients • Enrichment

1 Introduction

The numerical treatment of the diffusion equation

$$-\nabla \cdot (\kappa \nabla u) = f \quad \text{in } \Omega \subset \mathbb{R}^d$$
$$u = g_D \text{ on } \partial\Omega$$

with a strongly discontinuous scalar diffusion coefficient κ is an important and time-consuming task in many applications, e.g. groundwater flow [1], electromagnetics [12], fuel cell modelling [19], conductivity and heat transfer in composite materials, neutron transport and many more. In general the diffusion coefficient κ encodes material properties and is usually modelled as a positive piecewise constant function with substantial jumps across material interfaces. Thus, the gradient of the solution u is discontinuous across such material interfaces.

M.A. Schweitzer (✉)
Institut für Parallele und Verteilte Systeme, Universität Stuttgart, Universitätsstr. 38,
70569 Stuttgart, Germany
e-mail: marc.alexander.schweitzer@ipvs.uni-stuttgart.de

M. Griebel and M.A. Schweitzer (eds.), *Meshfree Methods for Partial Differential Equations VI*, Lecture Notes in Computational Science and Engineering 89,
DOI 10.1007/978-3-642-32979-1_6, © Springer-Verlag Berlin Heidelberg 2013

A classical finite element discretization of (1) yields a discrete linear system where the condition number of the respective stiffness matrix is dependent not only on the meshsize but also on the (discontinuous) coefficient. Thus, an efficient iterative solver for such linear systems must converge with a rate that is independent of the meshsize and that is robust with respect to (jumps in) the diffusion coefficient. A number of domain decomposition and (geometric) multigrid methods have been proposed over the years either as standalone solvers or as preconditioners for a Krylov subspace scheme.[1] However, a common assumption made in these approaches is that the material interfaces are geometrically resolved on the coarsest triangulation, i.e. the material interfaces are aligned with the faces of the mesh on the coarsest level. Therefore, the geometric complexity of the material interfaces determines the coarsest admissible mesh and the respective number of degrees of freedom on the coarsest level of the discretization. Obviously this assumption has a severe impact on the practical (pre-asymptotic) performance of the respective iterative solver since the overall computational costs may be dominated by the solution of the resulting coarselevel system.

The main reason for this assumption is that the discontinuities of the solution and its gradient can only be captured along mesh faces with classical finite elements due to the use of piecewise polynomial shape functions. In meshfree methods however arbitrary discontinuities can be treated efficiently by appropriate enrichment functions. Thus, the goal of this paper is to develop an efficient meshfree multilevel solver that is robust with respect to the jumps of the diffusion coefficient κ by replacing the geometric resolution of the discontinuities of κ by the use of simple enrichment functions which resolve the induced discontinuities of the gradient of the solution directly. Thereby, we can employ a much coarser discretization (in terms of degrees of freedom) than classical finite element solvers such that the overall computational cost is not dominated by the cost of the coarse level solve. Thus, the geometric complexity of the material interfaces should not affect the overall performance of our solver.

The remainder of this paper is organized as follows. In Sect. 2 we present the model problem discussed in this paper before we summarize the construction of our multilevel partition of unity method and the employed enrichment functions in Sect. 3. The results of our numerical experiments in two and three space dimensions are given in Sect. 4. The observed convergence rates clearly indicate that our simple enrichment approach allows for a much coarser approximation of our model problem (1) than a classical finite element approach while obtaining acceptable but not entirely robust convergence rates. Finally, we conclude with some remarks in Sect. 5.

---

[1]There exist more involved extensions of the multigrid approach which are not based on a geometric but on an operator- or matrix-dependent prolongation approach like the black box multigrid [6] or the algebraic multigrid [18] approach. These more involved techniques are specifically designed to be more robust than classical geometric multigrid schemes.

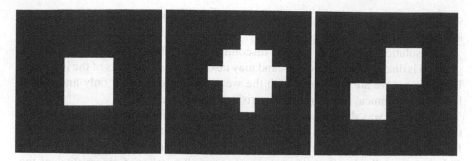

Fig. 1 Sketches of three reference material domain configurations employed in this paper. (*left*: a single square inclusion; *center*: a single inclusion with multiple reentrant corners; *right*: two square inclusions touching in a single point)

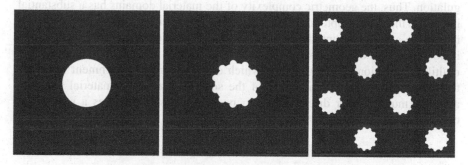

Fig. 2 Sketches of three reference material domain configurations employed in this paper. (*left*: a single spherical inclusion; *center*: a single inclusion with a more complex geometry; *right*: multiple disjoint inclusions with complex geometry)

2 Model Problem and Robust Solvers

Throughout this paper we are concerned with the scalar diffusion problem

$$
\begin{aligned}
-\nabla \cdot (\kappa \nabla u) &= f \quad \text{in } \Omega \subset \mathbb{R}^d \\
u &= g_D \quad \text{on } \Gamma_D \subset \partial\Omega \\
(\kappa \nabla u) \cdot n &= g_N \quad \text{on } \partial\Omega \setminus \Gamma_D
\end{aligned}
\tag{1}
$$

where the diffusion coefficient κ is a positive piecewise constant function. In particular we denote the pairwise disjoint open subsets $\Omega_p \subset \Omega$ on which κ is constant; i.e. $\kappa|_{\Omega_p} = \text{const}_p$, material domains. Moreover, we refer to $\Omega_0 \subset \Omega$ as the matrix material domain and $\Omega_p \subset \Omega$ with $p = 1, \ldots, m$ as inclusions, cmp Figs. 1 and 2.

Moreover, we assume that all inclusions are composed of the same material, i.e. $\kappa|_{\Omega_p} = \text{const}_p = \text{const}_1$ for $p = 2, \ldots, m$. Thus, the jump of the diffusion coefficient is given by $\frac{\text{const}_1}{\text{const}_0}$. The intersections $\bar{\Omega}_0 \cap \bar{\Omega}_p$ define the so-called material interface Γ, i.e. $\Gamma := \bigcup_{p=1}^m \bar{\Omega}_0 \cap \bar{\Omega}_p$.

The solution u of (1) is weakly discontinuous across the interface Γ, i.e. its gradient is discontinuous across Γ, and may also be singular at parts of the interface. In this paper we are concerned with the weak discontinuity of u only and do not focus on enrichment functions for the arising singularities.

In [5, 20] it was shown that classical geometric multigrid methods, even though they are not robust as a stand-alone solver, yield a robust and optimal preconditioner for the conjugate gradient method on quasi-uniform and bisection grids for (1). This is essentially due to the fact that only a small number of eigenvalues of the preconditioned system matrix are adversly affected by the jump of the diffusion coefficient $\frac{\text{const}_1}{\text{const}_0}$. Although the material domains are allowed to have complicated geometry in [5, 20] the interface Γ is assumed to be resoved by the coarsest triangulation. Thus, the geometric complexity of the material domains has a substantial impact on the pre-asymptotic performance of the respective preconditioner and may render the asymptotic optimality of the multigrid method practically useless, e.g. if no coarsening of the mesh is possible. The goal of this paper is to study if a multilevel partition of unity method which employs simple enrichment functions that encode the weak discontinuity of the solution across the material interface allows a much coarser discretization to be used yet still provides a robust and optimal preconditioner.

3 Multilevel Partition of Unity Method

The notion of a partition of unity method (PUM) was coined in [3, 4] and is based on the special finite element methods developed in [2]. The abstract ingredients of a PUM are

- A partition of unity (PU) $\{\varphi_i \mid i = 1, \ldots, N\}$ with

$$\varphi_i \in \mathscr{C}^r(\mathbb{R}^D, \mathbb{R}) \quad \text{and patches} \quad \omega_i := \text{supp}^\circ(\varphi_i),$$

- A collection of local approximation spaces

$$V_i(\omega_i, \mathbb{R}) := \text{span}\langle \vartheta_i^n \rangle \tag{2}$$

  defined on the patches ω_i for $i = 1, \ldots, N$.

With these two ingredients we define the PUM space

$$V^{PU} := \sum_{i=1}^N \varphi_i V_i = \text{span}\langle \varphi_i \vartheta_i^n \rangle; \tag{3}$$

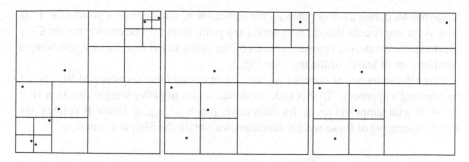

Fig. 3 Subdivision corresponding to a cover on level $J = 4$ with initial point cloud (*left*), derived coarser subdivisions on level 3 (*center*), and level 2 (*right*) with respective coarser point cloud

i.e., the shape functions of a PUM space are simply defined as the products of the PU functions φ_i and the local approximation functions ϑ_i^n. The PU functions provide the locality and global regularity of the product functions whereas the functions ϑ_i^n, i.e. the local spaces V_i, equip V^{PU} with its approximation power.

In general the local approximation space $V_i := \mathrm{span}\langle\vartheta_i^m\rangle$ associated with a particular patch ω_i of a PPUM space V^{PU} consists of two parts: A smooth approximation space, e.g. polynomials $\mathscr{P}^{p_i}(\omega_i) := \mathrm{span}\langle\psi_i^s\rangle$, and an enrichment part $\mathscr{E}_i(\omega_i) := \mathrm{span}\langle\eta_i^t\rangle$, i.e.

$$V_i(\omega_i) = \mathrm{span}\langle\vartheta_i^m\rangle = \mathscr{P}^{p_i}(\omega_i) + \mathscr{E}_i(\omega_i) = \mathrm{span}\langle\psi_i^s, \eta_i^t\rangle. \qquad (4)$$

The enrichment space $\mathscr{E}_i$ is a problem-dependent component of the local approximation space tailored to certain features of the solution which are known a priori and cannot be approximated well by polynomials e.g. discontinuities and singularities. The specific enrichment functions employed in this paper are given in Sect. 3.1.

The fundamental construction principle employed in [8,9,14] for the construction of a multilevel sequence of PUM spaces V_k^{PU}, i.e. a sequence of covers $C_{\Omega,k}$, is a d-binary tree. Based on the given point data $P = \{x_i \mid i = 1, \ldots, \hat{N}\}$, we subdivide a bounding-box $\mathscr{C}_\Omega \supset \Omega$ of the domain Ω until each cell

$$\mathscr{C}_i = \prod_{l=1}^{d}(c_i^l - h_i^l, c_i^l + h_i^l)$$

associated with a leaf of the tree contains at most a single point $x_i \in P$, see Fig. 3. We obtain an overlapping cover $C_{\Omega,J} := \{\omega_i\}$ with respect to the finest refinement level J from this tree by defining the cover patches ω_i by simple uniform and isotropic scaling

$$\omega_i := \prod_{l=1}^{d}(c_i^l - \alpha h_i^l, c_i^l + \alpha h_i^l), \quad \text{with } \alpha > 1. \qquad (5)$$

Note that we define a cover patch ω_i for leaf-cells $\mathscr{C}_i$ that contain a point $x_i \in P$ as well as for *empty* cells that do not contain any point from P. The coarser covers $C_{\Omega,k}$ are defined considering coarser versions of the constructed tree, i.e., by removing a complete set of leaves of the tree, see Fig. 3.

From the sequence of covers $C_{\Omega,k}$ we obtain a respective sequence of PUs $\{\varphi_{i,k}\}$ by Shepard's approach. To this end, we define a non-negative weight function $W_i :$ $\Omega \to \mathbb{R}$ with $\operatorname{supp}(W_i) = \omega_i$ for each cover patch ω_i, e.g. d-linear B-splines. By simple averaging of these weight functions we obtain the Shepard functions

$$\varphi_i(x) := \frac{W_i(x)}{S(x)}, \quad \text{with} \quad S(x) := \sum_{l=1}^{N} W_l(x). \tag{6}$$

Note that

$$S(x)|_{\omega_i} = \sum_{l \in C_i} W_l(x)$$

with $C_i := \{l \mid \omega_l \cap \omega_i \neq \emptyset\}$ and $\operatorname{card} C_i \ll N$. This approach yields a PU that satisfies the assumptions of the error analysis given in [4]. Moreover, the PU (6) based on a cover $C_\Omega = \{\omega_i\}$ obtained from the scaling of a tree decomposition (5) satisfies the flat-top property (for a particular range of $\alpha \in (1, 2)$), see [11].

Finally, we obtain a sequence of PUM spaces

$$V_k^{PU} := \sum_{i=1}^{N} \varphi_{i,k} V_{i,k} = \operatorname{span}\langle \varphi_{i,k} \vartheta_{i,k}^n \rangle;$$

from the sequences of PUs $\{\varphi_{i,k}\}$ and covers $C_{\Omega,k}$ by selecting appropriate local approximation spaces $V_{i,k} = \mathscr{P}^{p_{i,k}} + \mathscr{E}_{i,k}$ for each patch $\omega_{i,k} \in C_{\Omega,k}$. Note that our enrichment approach involves an additional local basis transformation [16] which constructs a locally stable basis for the enriched local approximation spaces V_i, see [16] for details. As transfer operators between these spaces we employ localized L^2-projections as prolongation operators $\Pi_k : V_{k-1}^{PU} \to V_k^{PU}$ and their adjoints Π_k^t as restriction operators, see [9, 14]. Observe that the construction of these transfer operators is straight forward for arbitrary enrichment functions.

3.1 Enrichment Functions

Recall that the solution u of our model problem (1) is weakly discontinuous across the interfaces where the diffusion coefficient κ is strongly discontinuous. Classical finite element methods resolve these interfaces geometrically by mesh refinement to capture this particular behavior of the solution. In our partition of unity approach the discontinuities of ∇u are explicitly incorporated into the approximation space V^{PU} of (3) via the local enrichment spaces $\mathscr{E}_i$ independent of the refinement of the cover c_Ω.

A very common choice of enrichment functions for the approximation of weak discontinuities across an interface Γ is based on the signed distance function $\eta_\Gamma(x) := \mathrm{sdist}(x, \Gamma)$ to the interface, see e.g. [7, 17] and the references therein.

Let us assume that the interface Γ is a closed curve and induces a disjoint decomposition of our domain Ω into Ω_+ and Ω_-. Moreover, let us assume that we employ linear polynomials in our local approximation spaces, i.e. $V_i = \mathscr{P}^1 + \mathscr{E}_i$, for the approximation of locally smooth solution components. Then it is very natural to look for local enrichment spaces $\mathscr{E}_i$ such that the enriched approximation space V^{PU} (3) can resolve functions that are globally continuous, piecewise linear in Ω, and linear in Ω_+ and Ω_-.

From a theoretical point of view, the optimal construction for the local approximation spaces $V_i = \mathscr{P}^{p_i} + \mathscr{E}_i$ would be based on a coordinate system that employs the normal to the interface (via the distance) to capture the kink and the (locally) tangential directions to describe smooth components. However, this is rather challenging in practice since the material interfaces may be non-linear. Thus, we limit ourselves to the use of an enrichment space $\mathscr{E}_i$ that captures the linear growth in normal direction to the interface only; i.e. we use the two dimensional enrichment space

$$\mathscr{E}_i := \mathrm{span}\langle \max(0, \eta_\Gamma), \min(0, \eta_\Gamma) \rangle = \mathrm{span}\langle \eta_\Gamma, |\eta_\Gamma| \rangle \qquad (7)$$

throughout this paper only. Moreover, we define the two subregions Ω_+ and Ω_- via the matrix material domain and the union of the inclusions, i.e.

$$\Omega_- := \Omega_0, \quad \Omega_+ := \bigcup_{i=1}^{m} \Omega_p,$$

such that the material interface is given by $\bar{\Omega}_+ \cap \bar{\Omega}_-$.

In general the construction of an exact signed distance function to an arbitrary interface Γ is not trivial and its evaluation may be rather expensive. Since we are concerned with the definition of enrichment functions that essentially have a piecewise linear character with respect to Ω_+ and Ω_- with a linear growth in normal direction to the interface Γ only, we use a simple approximation of the signed distance function to Γ by computational solid geometry operations. To this end, we assume that the material domains Ω_p with $p = 1, \ldots, m$ associated with the inclusions are unions of simple geometric objects, e.g. squares and spheres, for which analytic signed distance functions are available and define the enrichment functions by respective boolean operations. As an example let us consider the case of just two inclusions Ω_1 and Ω_2 given by

$$\Omega_1 := \{x \in \Omega : \|x - c_1\|_\infty < r_1\}, \quad \Omega_2 := \{x \in \Omega : \|x - c_2\|_2 < r_2\}.$$

For these inclusions the respective signed distance functions are obviously given by

$$\eta_1(x) := r_1 - \|x - c_1\|_\infty, \quad \eta_2(x) := r_2 - \|x - c_2\|_2$$

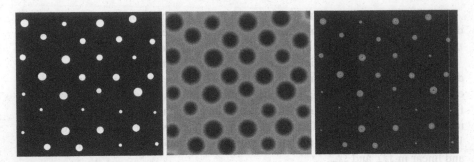

Fig. 4 Sketch of a randomized material domain configuration with 32 spherical inclusions (*left*) and the respective enrichment functions (*center*: matrix material domain, *right*: inclusions)

and we define our enrichment functions (7) for the two subdomains

$$\Omega_+ = \Omega_1 \cup \Omega_2, \quad \Omega_- = \Omega_0 = \Omega \setminus (\Omega_1 \cup \Omega_2)$$

simply as

$$\eta_+(x) := \max(0, \max(\eta_1(x), \eta_2(x))), \quad \eta_-(x) = \max(0, -\max(\eta_1(x), \eta_2(x))). \tag{8}$$

Thus, independent of the number of inclusions m we construct just two enrichment functions η_- and η_+, cmp Fig. 4. These enrichment functions are both employed on all patches ω_i which intersect the material interface Γ.

3.2 Essential Boundary Conditions

Throughout this paper we are concerned with the model problem (1) in two and three space dimensions and essential boundary conditions at least on part of the boundary $\partial\Omega$. In this paper we use Nitsche's method [10, 13, 15] for the implementation of the boundary conditions. Thus, the weak formulation of

$$-\nabla \cdot (\kappa \nabla u) = f \quad \text{in } \Omega \subset \mathbb{R}^d$$
$$u = g_D \quad \text{on } \partial\Omega$$

is given by

$$a_{C_\Omega,\tau}(u, v) = \langle l_{C_\Omega,\tau}, v \rangle$$

with the cover-dependent symmetric bilinearform

$$a_{C_\Omega,\tau}(u, v) := \int_\Omega (\kappa \nabla u) \cdot \nabla v \, dx - \int_{\partial\Omega} \left(((\kappa \nabla u) \cdot n)v + u((\kappa \nabla v) \cdot n) \right) ds$$
$$+ \tau \sum_{\omega_i \in C_{\partial\Omega}} \operatorname{diam}(\gamma_i)^{-1} \int_{\gamma_i} uv \, ds$$

and the associated linear form

$$\langle l_{C_\Omega, \tau}, v \rangle := \int_\Omega f v \, dx - \int_{\partial \Omega} g((\kappa \nabla v) \cdot n) \, ds + \tau \sum_{\omega_i \in C_{\partial \Omega}} \text{diam}(\gamma_i)^{-1} \int_{\gamma_i} g v \, ds$$

where the cover of the boundary is defined as

$$C_{\partial \Omega} := \{\omega_i \in C_\Omega \mid \gamma_i \neq \emptyset\}, \quad \text{and} \quad \gamma_i := \omega_i \cap \partial \Omega.$$

The regularization parameter τ is computed from a generalized eigenvalue problem [10].

4 Numerical Experiments

In the following we present the results of our numerical experiments with the proposed multilevel PUM. Here, we consider uniformly refined covers C_Ω with $\alpha = 1.2$ in (5) and employ linear polynomials as smooth local approximation spaces $\mathscr{P}^{p_i}$ only. The coarsest discretization level consists of just four patches and thus employs at most 20 degrees of freedom in two space dimensions; the three linear polynomials and the two enrichment functions (8) on each patch. The finest discretization level presented comprises over 12 million degrees of freedom in two space dimensions and over 8 million degrees of freedom in three space dimensions.

In all experiments we use a conjugate gradient method preconditioned by a $V(1, 1)$ multigrid cycle with block-Gauss-Seidel smoothing [9] for the solution of the symmetric positive definite linear system

$$A\tilde{u} = \hat{f}$$

arising from the Galerkin discretization of (1) using the sequence of PUM spaces presented in Sect. 3 with the enrichment functions (8). Moreover, we use $\Omega = (-1, 1)^d$ with $d = 2, 3$, $f = 0$ and the essential boundary conditions $g_D = 0$ at the boundary-segment with $x = -1$, $g_D = 1$ where $x = 1$ and homogeneous Neumann boundary conditions along the remaining boundary. We consider the material domain configurations depicted in Figs. 1, 2 and 4 in two space dimensions as well as the three-dimensional analogues to the configurations depicted in Fig. 1. In all experiments we fix the diffusion coefficient in the matrix material to 1, i.e. $\kappa|_{\Omega_-} = 1$, and vary the diffusion coefficient of the inclusions $\kappa|_{\Omega_+} = \epsilon \in \{10^{-6}, 10^{-4}, 10^{-2}, 1, 10^2, 10^4, 10^6\}$. Thus, we consider jumps of size ϵ.

As an initial guess for our solver we use $\tilde{u}_0 = 0$ and stop the iteration when the l^2-norm of the residual drops below 10^{-10}, i.e.

$$\|\hat{f} - A\tilde{u}_{iter}\|_{l^2} \leq 10^{-10}.$$

Table 1 Results of Example 1: number of $V(1, 1)$ preconditioned conjugate gradient iterations for a single square inclusion, cmp Fig. 1

Level	$\epsilon = 10^{-6}$	$\epsilon = 10^{-4}$	$\epsilon = 10^{-2}$	$\epsilon = 10^{0}$	$\epsilon = 10^{2}$	$\epsilon = 10^{4}$	$\epsilon = 10^{6}$
2	13	13	13	14	15	17	17
3	18	18	18	17	18	19	19
4	20	19	19	19	22	23	23
5	22	21	21	20	23	23	23
6	23	22	22	21	24	25	25
7	23	22	22	22	24	25	25
8	24	23	22	22	25	26	26
9	24	23	23	22	25	26	26
10	24	23	23	22	25	27	27
11	24	23	23	23	25	26	26

In the tables below we give the number of iterations necessary to satisfy this stopping criterion for the various considered material domain configurations.

Example 1 (Square inclusion, 2D). In our first simple example we are concerned with just a single inclusion $\Omega_1 = (-\frac{1}{3}, \frac{1}{3})^2$, see Fig. 1. This simple inclusion can of course be resolved easily on a coarse mesh but serves as a reference test case for further experiments.

When the material coefficient of Ω_1 is much smaller than that of the matrix material, i.e. $\epsilon \ll 1$, we are essentially dealing with diffusion within the inclusion only whereas the solution in the matrix material vanishes (asymptotically) and since Ω_1 is a simple convex square this should be rather easy to solve. On the other hand, if the diffusion coefficient within Ω_1 is large compared to κ on Ω_0 diffusion is essentially observed within Ω_0 which is non-convex only. Thus, the problems with $\epsilon \gg 1$ are in general more challenging.

A robust solver however must handle both cases with comparable efficiency. From the iteration counts given in Table 1 we can clearly observe that our multilevel preconditioner yields an optimal and very much robust solver. The number of iterations is essentially unaffected by the number of levels as well as by the size of the jump in the material properties; we find between 23 and 27 iterations are necessary to reduce the residual to the prescribed tolerance.

Example 2 (Star-like inclusion, 2D). As a second example we consider a single inclusion with multiple re-entrant corners, see Fig. 1 (center) so that a singular solution arises in the limit $\epsilon \to 0$ as well as $\epsilon \to \infty$. Moreover the inclusion is shifted from the center of the domain Ω to consider more arbitrary overlaps of the cover patches with the material interface. From the iteration counts summarized in Table 2 we find a small deterioration in the robustness. Here, we achieve a residual $<10^{-10}$ is about 30 iterations. For a fixed value of ϵ the number of iterations is essentially independent of the number of levels. If we fix the number levels and vary the jump size ϵ we find a small increase in the number of iterations from the minimum of 23 for $\epsilon = 1$ to 29 for $\epsilon = 10^6$ and 36 for $\epsilon = 10^{-6}$. We attribute

Table 2 Results of Example 2: number of $V(1, 1)$ preconditioned conjugate gradient iterations for a star-like inclusions, cmp Fig. 1

Level	$\epsilon = 10^{-6}$	$\epsilon = 10^{-4}$	$\epsilon = 10^{-2}$	$\epsilon = 10^{0}$	$\epsilon = 10^{2}$	$\epsilon = 10^{4}$	$\epsilon = 10^{6}$
2	14	14	14	14	15	15	16
3	18	18	18	17	18	19	20
4	23	22	22	19	21	22	22
5	26	24	23	20	22	23	23
6	30	28	25	21	24	25	25
7	30	28	25	22	24	26	27
8	30	29	26	22	25	27	31
9	32	31	27	22	26	28	29
10	34	30	27	22	26	28	29
11	36	31	27	23	27	29	29

Table 3 Results of Example 3: number of $V(1, 1)$ preconditioned conjugate gradient iterations for two touching square inclusions, cmp Fig. 1

Level	$\epsilon = 10^{-6}$	$\epsilon = 10^{-4}$	$\epsilon = 10^{-2}$	$\epsilon = 10^{0}$	$\epsilon = 10^{2}$	$\epsilon = 10^{4}$	$\epsilon = 10^{6}$
2	15	15	15	14	14	14	14
3	18	18	18	17	20	21	23
4	21	21	21	19	23	25	26
5	24	23	23	20	25	27	30
6	26	25	24	21	27	28	28
7	28	26	25	22	28	30	30
8	28	27	26	22	28	30	30
9	29	28	27	22	29	31	31
10	30	28	27	22	29	31	31
11	30	29	28	23	30	32	32

this deterioration of the robustness to the stronger singular character of the solution which is not accounted for by the enrichment functions. The absolute number of iterations however is still in a very much acceptable range.

Example 3 (Two touching square inclusions, 2D). The results obtained for a material domain configuration with two square inclusions touching at a single corner, see Fig. 1 are presented in Table 3. This is another typical reference problem for robust multilevel methods where the material domain configuration can of course be resolved on a coarse mesh.

The measured number of iterations in this example are given in Table 3. From these numbers we find that apart from an initial jump from 23 for $\epsilon = 10^{0}$ to about 30 for any other value of ϵ our solver is robust.

Example 4 (Spherical inclusion, 2D). In our fourth example we now consider a single spherical inclusion $\Omega_1 = \{x \in \Omega : \|x\|_2 < \frac{1}{3}\}$, thus the material interface Γ is a smooth curve which cannot be represented well on a coarse mesh. The number of iterations displayed in Table 4 are almost identical to those of the previous

Table 4 Results of Example 4: number of $V(1, 1)$ preconditioned conjugate gradient iterations for a single spherical inclusion, cmp Fig. 2

Level	$\epsilon = 10^{-6}$	$\epsilon = 10^{-4}$	$\epsilon = 10^{-2}$	$\epsilon = 10^0$	$\epsilon = 10^2$	$\epsilon = 10^4$	$\epsilon = 10^6$
2	13	13	13	14	16	17	18
3	17	17	17	17	20	21	20
4	20	20	19	19	21	23	23
5	22	22	20	20	22	24	24
6	24	23	21	21	24	25	25
7	25	24	22	22	24	26	26
8	27	24	22	22	24	26	27
9	27	25	22	22	24	26	26
10	27	25	23	22	24	26	26
11	26	25	23	23	25	26	26

Table 5 Results of Example 5: number of $V(1, 1)$ preconditioned conjugate gradient iterations for a gear wheel-like inclusion composed of 12 overlapping spherical inclusions, cmp Fig. 2

Level	$\epsilon = 10^{-6}$	$\epsilon = 10^{-4}$	$\epsilon = 10^{-2}$	$\epsilon = 10^0$	$\epsilon = 10^2$	$\epsilon = 10^4$	$\epsilon = 10^6$
2	12	12	12	14	16	17	18
3	18	17	17	17	19	20	20
4	20	19	19	19	21	22	23
5	22	21	21	20	23	24	30
6	32	24	22	21	23	25	25
7	28	26	23	22	24	26	27
8	33	29	23	22	25	26	27
9	30	28	24	22	25	27	27
10	31	29	25	22	25	27	30
11	32	29	25	23	25	27	32

example; i.e. the solver is optimal and robust, which indicates that the geometry of the inclusion has no substantial impact on the convergence properties of our solver.

Example 5 (Gear wheel-like inclusion, 2D). Next we consider a single non-convex inclusion that is composed of 12 overlapping spheres, see Fig. 2. Note that the resolution of the respective material interface Γ in this example clearly involves a rather fine mesh, whereas our coarsest discretization level consists of 20 degrees of freedom only. The attained iteration counts are collected in Table 5. Again, we find almost the some solver efficiency and robustness as in the previous examples even though the geometry of the material interface is far more complex. Here, we need no more than 32 iterations to solve the linear system to the prescribed accuracy.

Example 6 (Multiple gear wheel-like inclusions, 2D). Next we are concerned with a regular distribution of eight such non-convex material domains, see Fig. 2. Thus, the material interace Γ in this problem is composed of 96 spheres and the resolution of Γ would certainly require a mesh with far more than just 20 vertices.

The results obtained for this example are collected in Table 6. From the iteration counts given there we can clearly observe that we obtain very much equivalent

Table 6 Number of $V(1, 1)$ preconditioned conjugate gradient iterations for eight gear wheel-like inclusions, cmp Fig. 2

Level	$\epsilon = 10^{-6}$	$\epsilon = 10^{-4}$	$\epsilon = 10^{-2}$	$\epsilon = 10^0$	$\epsilon = 10^2$	$\epsilon = 10^4$	$\epsilon = 10^6$
2	14	14	14	14	12	13	13
3	17	17	17	17	17	18	18
4	18	18	18	19	20	22	24
5	20	19	19	20	21	24	25
6	22	22	22	21	24	26	38
7	47	25	23	22	25	27	28
8	29	26	24	22	25	28	29
9	38	31	25	22	26	28	29
10	31	29	25	22	26	28	29
11	32	29	26	23	26	28	31

results as in the previous example. For instance, on level 11 which employs over 12 million degrees of freedom we need between 23 and 32 iterations. However, there is also one configuration that shows a different behavior. For $\epsilon = 10^{-6}$ on level 7 we need 47 iterations while for $\epsilon = 10^{-4}$ the solver reaches the prescribed accuracy within 25 iterations. Moreover, the results on levels 6 and 8 for $\epsilon = 10^{-6}$ are again much better with just 22 and 29 iterations. We claim that this unsatisfactory effect is essentially caused by a numerical integration error.

On level 7 we find a number of patches which have an extremely small intersection with the inclusions. This by itself is not a problem for our approach [16] which is confirmed by the fact that for other values of ϵ we find optimal results. The observed increase in the number of iterations hence must be due to the size of the jump ϵ which essentially yields a different weighting of the entries in the stiffness matrix associated with the two enrichment functions (8). The respective integrals are not computed exactly but by numerical integration which is independent of the size of the jump in this experiment. Hence, the respective integration errors are weighted with different values of ϵ and thus the overall accuracy in the assembly is not independent of ϵ.

Example 7 (Multiple random spherical inclusions, 2D). In our last example in two space dimensions we consider a randomized distribution of 32 disjoint spherical inclusions of varying size, see Fig. 4. The measured number of iterations are given in Table 7. Again these results are in good agreement with those of the previous experiments. We find iteration counts of 26 for very small values of ϵ and 36 iterations for very large values of ϵ on level 11. Note that in this example a similar effect as in the previous example is observed on levels 4–6 for $\epsilon = 10^6$ where the iteration count rises to 44 but then again drops down to 36 on finer levels.

Example 8 (Cubical inclusion, 3D). In our first example in three space dimensions we consider a single inclusion $\Omega_1 = (-\frac{1}{3}, \frac{1}{3})^3$. Again this case essentially serves as a reference case for further experiments. The observed number of iterations are collected in Table 8.

Table 7 Number of $V(1,1)$ preconditioned conjugate gradient iterations for 32 randomized spherical inclusions, cmp Fig. 4

Level	$\epsilon = 10^{-6}$	$\epsilon = 10^{-4}$	$\epsilon = 10^{-2}$	$\epsilon = 10^{0}$	$\epsilon = 10^{2}$	$\epsilon = 10^{4}$	$\epsilon = 10^{6}$
2	14	14	14	14	12	13	24
3	17	17	16	17	16	19	21
4	18	19	17	19	23	32	40
5	19	18	18	20	26	37	43
6	19	19	20	21	26	34	44
7	22	21	21	22	27	35	38
8	24	22	22	22	28	35	39
9	25	23	23	22	28	35	36
10	26	24	23	22	29	35	36
11	26	25	23	23	29	35	36

Table 8 Number of $V(1,1)$ preconditioned conjugate gradient iterations for single cubical inclusion in three dimensions

Level	$\epsilon = 10^{-6}$	$\epsilon = 10^{-4}$	$\epsilon = 10^{-2}$	$\epsilon = 10^{0}$	$\epsilon = 10^{2}$	$\epsilon = 10^{4}$	$\epsilon = 10^{6}$
2	10	10	10	10	15	21	23
3	11	11	11	11	15	17	18
4	11	11	11	11	15	17	18
5	12	12	12	11	14	16	17
6	11	11	11	11	15	17	17
7	12	12	12	11	14	15	15

Note that the number of iterations for the smooth solution case ($\epsilon = 10^{0}$) are substantially smaller than in our previous examples in two space dimensions. This is due to the use of the scaling factor $\alpha = 1.5$ in the definition of the cover patches (5) in this experiment whereas we used $\alpha = 1.2$ in the previous examples which essentially impacts the effectiveness of the smoother.

The convergence behavior of our solver is again independent of the number of levels for all values of ϵ and only slightly dependent on ϵ. For large values of ϵ we find an increase in the number of iterations to 15 compared to the minimum of 11 iterations for $\epsilon = 10^{0}$.

Example 9 (Two cubical inclusions touching along an edge, 3D). The results obtained for two cubical inclusions that touch along a complete edge; i.e. $\Omega_1 = (-\frac{2}{3}, 0)^2 \times (-\frac{1}{3}, \frac{1}{3})$ and $\Omega_2 = (0, \frac{2}{3})^2 \times (-\frac{1}{3}, \frac{1}{3})$, are given in Table 9. This is essentially a cylindrical version of the two-dimensional Example 3 and thus we expect to find similar results to those summarized in Table 3. From the numbers given in Table 9 we find that the scenarios with large ϵ are more challenging than those with small ϵ. The relative increase in the number of iterations, i.e., the deterioration in robustness, for large ϵ is somewhat larger than in two dimensions, yet still it is in an acceptable range for practice. The largest system on level 7 which employs over 8 million degrees of freedom can be solved up to the prescribed accuracy within 22 iterations. Moreover, the absolute size of the jump does not seem

Table 9 Number of $V(1, 1)$ preconditioned conjugate gradient iterations for two cubical inclusions touching along an edge in three dimensions

Level	$\epsilon = 10^{-6}$	$\epsilon = 10^{-4}$	$\epsilon = 10^{-2}$	$\epsilon = 10^0$	$\epsilon = 10^2$	$\epsilon = 10^4$	$\epsilon = 10^6$
2	12	12	12	10	16	19	20
3	12	12	12	11	17	21	22
4	13	13	12	11	16	21	22
5	13	13	13	11	17	21	22
6	13	13	13	11	17	21	22
7	14	13	13	11	17	22	23

Table 10 Number of $V(1, 1)$ preconditioned conjugate gradient iterations for two cubical inclusions touching in a single point in three dimensions

Level	$\epsilon = 10^{-6}$	$\epsilon = 10^{-4}$	$\epsilon = 10^{-2}$	$\epsilon = 10^0$	$\epsilon = 10^2$	$\epsilon = 10^4$	$\epsilon = 10^6$
2	13	13	13	10	17	23	24
3	12	12	12	11	18	31	37
4	12	12	12	11	18	32	41
5	12	12	12	11	18	33	42
6	13	12	12	11	18	32	42
7	13	13	12	11	18	32	42

to have a large influence on the number of iterations. The results for $\epsilon = 10^4$ and $\epsilon = 10^6$ are almost identical with roughly 20 iterations.

Example 10 (Two cubical inclusions touching in a single point, 3D). In our final experiment we consider two inclusions touching in a single point in three space dimensions; i.e. the three-dimensional analogue of Example 3. The number of iterations obtained for this example are given in Table 10. From these numbers we learn a very different behavior of our solver. There is still no dependence on the number of levels but the number of iterations is obviously dependent on the size of the jump. We need 42 iterations for $\epsilon = 10^6$, 32 iterations for $\epsilon = 10^4$ and 18 iterations for $\epsilon = 10^2$ compared to the optimal value of 11 iterations for smaller values of ϵ. Thus, our simple enrichment approach does not yield a robust solver in this experiment where the solution exhibits strong singularities along the edges of the touching material domains.

5 Concluding Remarks

In this paper we studied a multilevel partition of unity method with very simple enrichment functions for the efficient and robust solution of a scalar diffusion problem with substantial jumps in the diffusion coefficient and geometrically complex material interfaces. The main focus of our study was if the use of enrichment functions allows to use a much coarser discretization level than is possible with classical finite elements so that the optimality and robustness of our solver is attained already in the pre-assymptotic range. The presented numerical results in two dimensions indicate that the solver is virtually unaffected by the geometric complexity of the material interface. Thus, the presented approach to attain an optimal and robust solver is viable and promising. The results obtained in three dimensions, however, indicate that our enrichment functions may be too simple to deal with problems with arbitrary singularities. Moreover, we observed a deterioration of the convergence rate when the intersections of the material domains with the cover patches are very small. These stability issues and the use of more involved enrichment functions that can cope with arbitrary singularities are subject of current research.

References

1. R.E. Alcouffe, A. Brandt, J.E. Dendy Jr., J.W. Painter, The multi-grid methods for the diffusion equation with strongly discontinuous coefficients. SIAM J. Sci. Stat. Comput. **2**(4), 430–454 (1981)
2. I. Babuška, G. Caloz, J.E. Osborn, Special finite element methods for a class of second order elliptic problems with rough coefficients. SIAM J. Numer. Anal. **31**, 945–981 (1994)
3. I. Babuška, J.M. Melenk, The partition of unity finite element method: basic theory and applications. Comput. Meth. Appl. Mech. Eng. **139**, 289–314 (1996). Special issue on meshless methods
4. I. Babuška, J.M. Melenk, The partition of unity method. Int. J. Numer. Method Eng. **40**, 727–758 (1997)
5. L. Chen, M. Holst, J. Xu, Y. Zhu, Local multilevel preconditioners for elliptic equations with jump coefficients on bisection grids (2010). Preprint, arXiv:1006.3277v2
6. J.E. Dendy, Black box multigrid. J. Comput. Phys. **48**, 366–386 (1982)
7. T.-P. Fries, T. Belytschko, The extended/generalized finite element method: an overview of the method and its applications. Int. J. Numer. Method Eng. **84**(3), 253–304 (2010)
8. M. Griebel, M.A. Schweitzer, A particle-partition of unity method—part II: efficient cover construction and reliable integration. SIAM J. Sci. Comput. **23**(5), 1655–1682 (2002)
9. M. Griebel, M.A. Schweitzer, A particle-partition of unity method—part III: a multilevel solver. SIAM J. Sci. Comput. **24**(2), 377–409 (2002)
10. M. Griebel, M.A. Schweitzer, A particle-partition of unity method—part V: boundary conditions, in *Geometric Analysis and Nonlinear Partial Differential Equations*, ed. by S. Hildebrandt, H. Karcher (Springer, Berlin, 2002), pp. 517–540
11. M. Griebel, M.A. Schweitzer, A particle-partition of unity method—part VII: adaptivity, in *Meshfree Methods for Partial Differential Equations III*, ed. by M. Griebel, M.A. Schweitzer. Lecture Notes in Computational Science and Engineering, vol. 57 (Springer, Berlin, 2006), pp. 121–148
12. B. Heise, M. Kuhn, Parallel solvers for linear and nonlinear exterior magnetic field problems based upon coupled fe/be formulations. Computing **56**(3), 237–258 (1996)
13. J. Nitsche, Über ein Variationsprinzip zur Lösung von Dirichlet-Problemen bei Verwendung von Teilräumen, die keinen Randbedingungen unterworfen sind. Abh. Math. Sem. Univ. Hamburg **36**, 9–15 (1970–1971)
14. M.A. Schweitzer, *A Parallel Multilevel Partition of Unity Method for Elliptic Partial Differential Equations*. Lecture Notes in Computational Science and Engineering, vol. 29 (Springer, Berlin, 2003)
15. M.A. Schweitzer, An algebraic treatment of essential boundary conditions in the particle–partition of unity method. SIAM J. Sci. Comput. **31**, 1581–1602 (2009)

16. M.A. Schweitzer, Stable enrichment and local preconditioning in the particle–partition of unity method. Numer. Math. **118**(1), 137–170 (2011)
17. M.A. Schweitzer, Generalizations of the finite element method. Cent. Eur. J. Math. **10**(1), 3–24 (2012)
18. U. Trottenberg, C.W. Osterlee, A. Schüller, Multigrid, in *Appendix A: An Introduction to Algebraic Multigrid by* K. STÜBEN (Academic, San Diego, 2001), pp. 413–532
19. C. Wang, Fundamental models for fuel cell engineering. Chem. Rev. **104**, 4727–4766 (2004)
20. J. Xu, Y. Zhu, Uniform convergent multigrid methods for elliptic problems with strongly discontinuous coefficients. Math. Models Methods Appl. Sci. **18**(1), 77–105 (2008)

HOLMES: Convergent Meshfree Approximation Schemes of Arbitrary Order and Smoothness

Agustín Bompadre, Luigi E. Perotti, Christian J. Cyron, and Michael Ortiz

Abstract Local Maximum-Entropy (LME) approximation schemes are meshfree approximation schemes that satisfy consistency conditions of order 1, i.e., they approximate affine functions exactly. In addition, LME approximation schemes converge in the Sobolev space $W^{1,p}$, i.e., they are C^0-continuous in the conventional terminology of finite-element interpolation. Here we present a generalization of the Local Max-Ent approximation schemes that are consistent to arbitrary order, i.e., interpolate polynomials of arbitrary degree exactly, and which converge in $W^{k,p}$, i.e., they are C^k-continuous to arbitrary order k. We refer to these approximation schemes as High Order Local Maximum-Entropy Approximation Schemes (HOLMES). We prove uniform error bounds for the HOLMES approximates and their derivatives up to order k. Moreover, we show that the HOLMES of order k is dense in the Sobolev Space $W^{k,p}$, for any $1 \leq p < \infty$. The good performance of HOLMES relative to other meshfree schemes in selected test cases is also critically appraised.

Keywords High-order meshfree interpolation

A. Bompadre · L.E. Perotti · M. Ortiz (✉)
Graduate Aerospace Laboratories, California Institute of Technology, Pasadena, CA 91125, USA
e-mail: abompadr@alum.mit.edu; luigiemp@caltech.edu; ortiz@aero.caltech.edu

C.J. Cyron
Institute for Computational Mechanics, Technische Universität München, 85748 Garching, Germany
e-mail: cyron@lnm.mw.tum.de

M. Griebel and M.A. Schweitzer (eds.), *Meshfree Methods for Partial Differential Equations VI*, Lecture Notes in Computational Science and Engineering 89, DOI 10.1007/978-3-642-32979-1_7, © Springer-Verlag Berlin Heidelberg 2013

1 Introduction

Local Maximum-Entropy (LME) approximations schemes are a meshfree approximation method introduced in [1] (see also [2, 21]). This method combines, by means of a single parameter, shape functions of least width and maximum-entropy statistical inference. The LME are *convex approximation schemes*, in the sense that their shape functions are non-negative, and satisfy the consistency constrains up to order 1. In particular, they approximate affine functions exactly, and satisfy a weak Kronecker delta condition, which simplifies the imposition of boundary conditions.

The LME approximates can be used to solve PDEs numerically, in the style of meshfree Galerkin methods, see, e.g., [12]. This was done, in particular, in [1], where the LME approximates gave better results than finite element methods. An optimal transportation meshfree (OTM) method for simulating general solid and fluid flows that uses the LME approximates was developed in [14]. An extension of LME to analyze thin shells was presented in [16]. Related approximation methods include the Moving Least Squares (MLS) methods [15, 17], and radial basis functions [23]. A study of the convergence properties of LME was carried out in [5]. In that paper, it was shown that the error incurred by the LME approximates is quadratic on h, a measure of the density of the point set on the domain. Moreover, the error of the derivatives of the LME approximates is linear on h. As a result, the LME approximates are dense in $W^{1,q}$, for any $1 \leq q < \infty$. In the conventional terminology of finite-element interpolation, LME approximation schemes are 'C^0-continuous' and, as such, can only be applied to the solution of second-order problems.

A natural generalization of the LME approximation scheme is to extend them to satisfy higher-order consistency constraints. As already pointed out in [1], simply adding second-order constraints to the current set of constraints may lead to an infeasible method, and therefore some care is necessary to accomplish this task. Recently, this was done in [8, 11] by relaxing some of the constraints. We present a generalization of the LME approximation scheme to high-order consistency constraints, which we refer to as High-Order Local Maximum-Entropy Approximation Schemes (HOLMES) [4]. The shape functions of the HOLMES of order n satisfy the consistency constraints up to order n, i.e., they interpolate exactly all polynomials of degree less or equal to n. In order to guarantee feasibility, their nonnegativity constraint is eliminated. A drawback of not enforcing non-negativity of the shape functions is that, unlike LME, HOLMES does not satisfy a weak Kronecker delta property. As a result, the imposition of boundary conditions for HOLMES is somewhat more involved than in the LME case. However, imposing boundary conditions can be accomplished by using standard Lagrangian multipliers, as in [7, 12].

We analyze the theoretical convergence of the approximates constructed from the HOLMES scheme of order n. We prove that, for each $k \leq n$, the error of the derivatives of order k of the HOLMES approximates is of the order $O(h^{n+1-k})$, where h is a measure of the density of the point set. As a result, the HOLMES approximates of order n are dense in $W^{n,q}$, for any $1 \leq q < \infty$. Again, in the conventional parlance of finite-element interpolation, HOLMES is 'C^n-continuous' and can thus be applied to problems of arbitrary order $2(n+1)$. For instance, second-order HOLMES can be applied to problems concerned with plates and shells. We compare the performance of second-order HOLMES with that of other solvers on a number of benchmark problems. The numerical tests show that HOLMES is indeed comparable to other methods commonly used in practice.

The rest of the paper is organized as follows. In Sect. 2, we establish the notation and some basic definitions that are used throughout the paper. The HOLMES are defined in Sect. 3, and its convergence properties are stated in Sect. 4. An analysis of the performance of the second-order HOLMES in selected test cases is finally presented in Sect. 5.

2 Prolegomena

For $p > 0$, the p-norm of a vector $x = (x_1, \ldots, x_d) \in \mathbb{R}^d$ is equal to $|x|_p = \left(\sum_{i=1}^d |x_i|^p\right)^{1/p}$. The open d-ball $B_p(x, \delta)$ of radius δ centered at x is the set $\{y \in \mathbb{R}^d : |y - x|_p < \delta\}$.

By a *domain* Ω we understand a bounded subset of $\mathbb{R}^d$. Given a *point set* $P \in (\mathbb{R}^d)^N$, we denote by $\overline{\text{conv}}(P)$ its closed convex hull [19].

A *multiindex* α is a vector in $\mathbb{N}^d$, for some dimension d. The *order* of α is equal to $|\alpha| = \sum_{i=1}^d \alpha_i$. Given a point $x \in \mathbb{R}^d$, we denote by x^α the expression $\prod_{i=1}^d x_i^{\alpha_i}$. We assume the set of multiindices $\mathbb{N}^d$ has a fixed complete ordering. In particular, a vector x in $\mathbb{R}^{D_n}$, where $D_n = \binom{d+n}{n}$, can be written as $(x_\alpha)_{\alpha \in \mathbb{N}^d, |\alpha| < n}$.

Definitions 1 and 2 below formalize the notion of a point set P that approximates a domain Ω uniformly.

Definition 1 (*h-covering of order n*). We say that a point set P is an *h-covering* of Ω of order n with constant κ, where $h, \kappa > 0$, and $n \geq 1$ is an integer, if for every $x \in \Omega$ there exists P_x, a subset of P with $D_n = \binom{d+n}{n}$ node points at distance at most h of x, and such that the following condition is satisfied

$$\inf_{\substack{u \in \mathbb{R}^{D_n}, \ x_a \in P_x \\ |u|=1}} \max \left| \sum_{\alpha \in \mathbb{N}^d : |\alpha| \leq n} h^{-|\alpha|} u_\alpha (x - x_a)^\alpha \right| \geq \kappa. \tag{1}$$

The inf-sup condition (1) measures how close the following matrix is to being singular

$$\left(h^{-|\alpha|} (x - x_a)^\alpha \right)_{\substack{a \in P_x, \\ \alpha \in \mathbb{N}^d, |\alpha| \leq n}}. \tag{2}$$

We mention that a sufficient condition for the inf-sup condition to hold is the existence, for each point $x \in \Omega$, of a principal lattice of order n included in P, with diameter h and aspect ratio σ containing x. Such a condition is standard in Finite Element Methods of order n, see, e.g., [6].

Definition 2 (*h*-density). We say that a point set $P \subset \mathbb{R}^d$ has *h-density* bounded by $\tau > 0$ if for every $x \in \mathbb{R}^d$, $\#\big(P \cap \bar{B}_p(x,h)\big) \leq \tau$.

For the High-Order Local Max-Ent Approximation Schemes we define in Sect. 3, we will assume the corresponding point set P to be an *h*-covering of order n with constant κ, and to have *h*-density bounded by τ. In particular, we will be interested on sequences P_k with the following properties.

Definition 3 (Strongly regular sequence of point sets). Let $\Omega \subset \mathbb{R}^d$ be a domain. Let P_k be a sequence of point sets of Ω. We say that P_k is a *strongly regular* sequence of point sets of order n if there is a sequence of positive real numbers $h_k \to 0$ and parameters $\kappa, \tau > 0$ such that

(i) For every k, P_k is an h_k-covering of Ω of order n and constant κ.
(ii) For every k, the node set P_k has h_k-density bounded by τ.

3 High-Order Local Maximum-Entropy Approximation Schemes

In this section we define the High-Order Local Maximum-Entropy approximation Schemes (HOLMES), and discuss some of their features. We start by reviewing the LME approximation schemes.

3.1 Local Maximum-Entropy Approximation Schemes

The Local Maximum Entropy (LME) Approximation Schemes were introduced in [1] as a method for approximating functions over a convex domain. The shape functions of the LME are nonnegative, and satisfy the consistency conditions of order 1. The LME approximation scheme is a convex approximation scheme that aims to satisfy two objectives simultaneously:

1. *Unbiased statistical inference* based on the nodal data.
2. Shape functions of *least width*.

Since for each point x, the shape functions of a convex approximation scheme are nonnegative and add up to 1, they can be thought of as the probability distribution of a random variable. The statistical inference of the shape functions is then measured by the *entropy* of the associated probability distribution, as defined in information theory [13]. The entropy of a probability distribution p over I is:

$$H(p) = -\sum_{a \in I} p_a \log p_a, \tag{3}$$

where $0 \log 0 = 0$. The least biased probability distribution p is that which maximizes the entropy. In addition, the *width* of a non-negative function w about a point ξ is identified with the second moment $U_\xi(w) = \int_\Omega w(x)|x - \xi|^2 \, dx$. Thus, the width $U_\xi(w)$ measures how concentrated w is about ξ. According to this measure, the most local shape functions are the ones that minimize the total width [18]

$$U(W) = \sum_{a \in I} U_a(w_a) = \int_\Omega \sum_{a \in I} w_a(x)|x - x_a|^2 \, dx. \qquad (4)$$

The Local Maximum-Entropy approximation schemes combine the functionals (3) and (4) into a single objective. More precisely, for a parameter $\beta > 0$, the LME approximation scheme is the minimizer of the functional

$$F_\beta(W) = \beta U(W) - H(W) \qquad (5)$$

under the restriction of first-order consistency and nonnegativity of the shape functions. We mention that the minimizer of (5) can be computed pointwise through a convex optimization problem.

3.2 Definition of High-Order Local Maximum Entropy Approximation Schemes

As already pointed out in [1], simply adding the second-order consistency conditions to the set of constrains of LME leads to an infeasible problem. Therefore, in order to extend the LME schemes to higher-order constraints, some of the constrains have to be relaxed. In [8], a Second order Maximum Entropy (SME) approximation scheme was introduced by enforcing the second-order constraints to be equal to a nonnegative *gap function*. Alternative higher-order extensions of LME have been presented in [11, 21].

In this section, we present a high-order generalization of the LME approximation schemes, which we name High-Order Local Maximum-Entropy Approximation Schemes (HOLMES). For an integer $n \geq 0$, the HOLMES of order n enforce the consistency conditions up to order n, and jettison the nonnegative constraint on the shape function. The shape functions $w_a(x)$, $a \in I$, can be written as $w_a(x) = w_a^+(x) - w_a^-(x)$, where $w_a^+(x)$ and $w_a^-(x)$ are nonnegative. We define the following *modified* entropy measure.

$$H(w^+(x), w^-(x)) = -\sum_{a \in I} w_a^+(x) \log w_a^+(x) - \sum_{a \in I} w_a^-(x) \log w_a^-(x). \qquad (6)$$

The total width of w^+, w^- with respect to a p-norm is as follows.

$$U_p(w^+, w^-) = \int_\Omega \sum_{a \in I} \left(w_a^+(x) + w_a^-(x)\right) |x - x_a|_p^p \, dx. \tag{7}$$

Note that $U_p(w^+, w^-)$ can be minimized pointwise. More precisely, the minimizers of U_p also minimize the function

$$U_p(w^+(x), w^-(x)) = \sum_{a \in I} \left(w_a^+(x) + w_a^-(x)\right) |x - x_a|_p^p, \tag{8}$$

almost everywhere. Let $\gamma, h > 0$ be two parameters. As in LME, the shape functions of HOLMES are chosen to minimize a combination of the objective functions (6) and (8), namely, for each point x,

$$\min f_\gamma(x, w^+(x), w^-(x)) = \gamma h^{-p} U_p(w^+(x), w^-(x)) - H(w^+(x), w^-(x)), \tag{9}$$

subject to the consistency constrains of order n. An equivalent mathematical formulation of the HOLMES of order n on a point x is as follows

$$\min f_\gamma(x, w^+(x), w^-(x)) = \gamma h^{-p} \sum_{a \in I} \left(w_a^+(x) + w_a^-(x)\right) \cdot |x - x_a|_p^p$$

$$+ \sum_{a \in I} w_a^+(x) \log w_a^+(x) + \sum_{a \in I} w_a^-(x) \log w_a^-(x),$$

subject to:

$$w_a^+(x), w_a^-(x) \geq 0, \ a \in I,$$

$$\sum_{a \in I} \left(w_a^+(x) - w_a^-(x)\right) = 1,$$

$$\sum_{a \in I} \left(w_a^+(x) - w_a^-(x)\right) h^{-|\alpha|}(x - x_a)^\alpha = 0, \ \forall \alpha \in \mathbb{N}^d, 1 \leq |\alpha| \leq n. \tag{10}$$

Note that, in this formulation, h is used as a scaling factor in the objective function as well as in the consistency constrains. Clearly, the problem can be equivalently formulated without h. However, the scaling h conveniently simplifies the analysis of the Lagrangian multipliers that follows.

We note that, when $n = 1$ and $p = 2$, the HOLMES approximation scheme is similar to (but not equal to) the LME approximation scheme of [1] of parameter $\beta = \gamma / h^2$. One difference in the formulation of both optimization problems is that the HOLMES schemes allow negative shape functions. Consequently, the shape functions do not define a probability distribution on each point x. In turn, $H(w^+, w^-)$ is no longer the entropy of a probability distribution. In HOLMES, the term $H(w^+, w^-)$ of the objective function penalizes values of w^+ and of w^- greater than one. Moreover, as in LME, the resulting shape functions decrease exponentially, as shown in [4].

The next Proposition shows sufficient conditions for feasibility of (10).

Proposition 1. *Let $P = \{x_a : a \in I\}$ be a nodal set that is an h-covering of Ω of order n and constant $\kappa > 0$. Then, for every point $x \in \Omega$, Problem (10) is feasible. Moreover, there exist shape functions $w_{++}^+, w_{++}^- \in \mathbb{R}_{++}^{|I|}$ that satisfy the constraints of (10) for x.*

3.3   Dual Formulation of HOLMES

We make use of duality theory in convex optimization (cf. [19]) to characterize the optimal solution of (10) of order n. For each multiindex $\alpha \in \mathbb{N}^d$, $|\alpha| \leq n$, let $\lambda_\alpha \in \mathbb{R}$ be a variable. Let $\lambda \in \mathbb{R}^{D_n}$ be the vector containing all the variables λ_α, namely, $\lambda = (\lambda_\alpha)_{\alpha \in \mathbb{N}^d, |\alpha| \leq n}$. For each node point $x_a \in P$, we define the functions $f_{n,a}^+$, $f_{n,a}^- : \mathbb{R}^d \times \mathbb{R}^{D_n} \to \mathbb{R}$ as follows.

$$f_{n,a}^+(x, \lambda) = -1 - h^{-p}\gamma|x - x_a|_p^p - \left(\sum_{\alpha \in \mathbb{N}^d, |\alpha| \leq n} h^{-|\alpha|}\lambda_\alpha(x - x_a)^\alpha\right), \quad (11\text{a})$$

$$f_{n,a}^-(x, \lambda) = -1 - h^{-p}\gamma|x - x_a|_p^p + \left(\sum_{\alpha \in \mathbb{N}^d, |\alpha| \leq n} h^{-|\alpha|}\lambda_\alpha(x - x_a)^\alpha\right). \quad (11\text{b})$$

Moreover, we define the functions $w_a^+, w_a^- : \mathbb{R}^d \times \mathbb{R}^{D_n} \to \mathbb{R}$ as follows.

$$w_a^+(x, \lambda) = \exp[f_{n,a}^+(x, \lambda)], \qquad w_a^-(x, \lambda) = \exp[f_{n,a}^-(x, \lambda)]. \quad (12)$$

We define the *partition function* $Z_n : \mathbb{R}^d \times \mathbb{R}^{D_n} \to \mathbb{R}$,

$$Z_n(x, \lambda) = \lambda_0 + \sum_{a \in I}\left(w_a^+(x, \lambda) + w_a^-(x, \lambda)\right). \quad (13)$$

The optimal solution of (10) can be described in terms of the unique minimizer of the partition function, as the next proposition shows.

Proposition 2. *Let P be a node set, and assume it is an h-covering of order n with constant κ of Ω. Then, the optimal solution of (10) for $x \in \Omega$ is*

$$w_a^{+*}(x) = w_a^+\left(x, \lambda^*(x)\right), \, a \in I, \quad (14\text{a})$$

$$w_a^{-*}(x) = w_a^-\left(x, \lambda^*(x)\right), \, a \in I, \quad (14\text{b})$$

where

$$\lambda^*(x) = \operatorname{argmin}_{\lambda \in \mathbb{R}^{D_n}} Z_n(x, \lambda). \quad (15)$$

Moreover, the minimizer $\lambda^*(x)$ is unique.

The smoothness of the shape functions depends on the parameter p chosen, as the next proposition shows.

Proposition 3. *Assume the hypothesis of Proposition 2 holds. Then, the shape functions of HOLMES, w_a^{+*} and w_a^{-*}, $a \in I$, are $C^\infty(\Omega)$ if p is even, and $C^{p-1}(\Omega)$ if p is odd.*

Since throughout this paper we assume that $p \geq n + 1$, the HOLMES shape functions are C^n functions.

4 Convergence of HOLMES

By the sufficient conditions for convergence of general approximation schemes given in [5], the following theorems hold. We leave the proofs for the extended version of this paper [4].

Theorem 1 (Uniform interpolation error bound of HOLMES). *Let $\{I_k, W_k\}$ be a HOLMES approximation scheme of order $n \geq 0$, and P_k a corresponding family of point sets. Suppose that P_k is a strongly regular sequence of order n for a sequence h_k. Let $\ell \geq 0$ be an integer. Let $m = \min\{n, \ell\}$. Then, there exists a constant $C < \infty$ such that*

$$|D^\alpha u_k(x) - D^\alpha u(x)| \leq C \left\| D^{m+1} u \right\|_\infty h_k^{m+1-|\alpha|}, \tag{16}$$

for every k, $u \in C^{\ell+1}(\overline{\mathrm{conv}}(\Omega))$, $|\alpha| \leq m$, and $x \in \Omega$.

Theorem 2. *Let Ω be a domain satisfying the segment condition. Suppose that the assumptions of Theorem 1 hold and that $h_k \to 0$. Let $1 \leq q < \infty$. Then, for every $u \in W^{n,q}(\Omega)$ there exists a sequence $u_k \in X_k$ such that $u_k \to u$.*

We note that other meshfree methods satisfy similar error bounds, notably Moving Least Squares (MLS) [3, 10], and Moving Least-Square Reproducing Kernel (MLSRK) methods [15].

5 Examples

In this section we study the convergence and accuracy of second order HOLMES functions. To this end we compare the performance of these basis functions with second order B-splines, moving least squares (MLS) functions, and C^0 Lagrange finite elements of degree 2 (FEM). For the MLS functions third order B-splines are used as window functions. All examples presented in this section make use of a uniform discretization with characteristic discretization length h along each coordinate axis, and Gauss quadrature subordinate to an auxiliary triangulation is used for

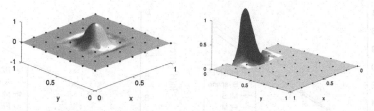

Fig. 1 Examples of 2nd order HOLMES shape functions

numerical integration. In case of B-splines both the nodes in the physical space and the knots in the parameter space are uniformly distributed. Numerical quadrature is performed in d dimensions for FEM basis functions with n_{GP}^d Gauss points per finite element. For B-splines Gauss integration with n_{GP}^d Gauss points is performed in the parameter space on the intervals between adjacent knots and for HOLMES and MLS functions with n_{GP}^d Gauss points on the intervals between adjacent nodes in the physical space. Boundary conditions on derivatives are imposed by Lagrange multipliers as described in [7] wherever B-splines were used. For HOLMES and MLS functions, Lagrange multipliers are used to impose boundary conditions both on function values and derivatives following the collocation point method (see Sect. 3.3.2, Chap. 10 of [12]). According to the collocation point method, the interpolation functions for Lagrange multipliers are Dirac delta functions centered at a set of collocation points where the boundary conditions are imposed exactly. In all our examples the collocation points correspond to the discretization nodes on the Dirichlet boundary of the domain.

The strict support of the HOLMES basis functions in d dimensions spans the entire space $\mathbb{R}^d$. In practice, their amplitude decreases rapidly with increasing distance from their respective nodes, see Fig. 1. By virtue of this decay, the basis functions may be truncated within numerical precision at an l_p-distance $r_{p,HOLMES}$. This range can be estimated as $r_{p,HOLMES} \approx \left(-\frac{\ln \epsilon + 1}{\gamma}\right)^{\frac{1}{p}}$, where ϵ represents the truncation tolerance. In the computation of the basis functions at a point x in the domain, the optimization problem (10) may then be solved neglecting the influence of all nodes at an l_p-distance larger than $r_{p,HOLMES}$.

5.1 Elastic Rod Under Sine Load

The static displacements $u(x)$ of a linearly elastic rod with unit material and geometry parameters and fixed ends under sine load are governed by the partial differential equation

$$u_{xx}(x) = \sin(2\pi x),$$
$$x \in [0; 1], u(0) = u(1) = 0. \qquad (17)$$

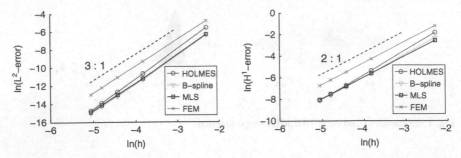

Fig. 2 L^2- and H^1-error convergence of the displacement field of an elastic rod under sine load approximated with different types of basis functions: higher order max-ent functions (HOLMES), B-splines, moving least squares functions (MLS) and Lagrange polynomials (FEM)

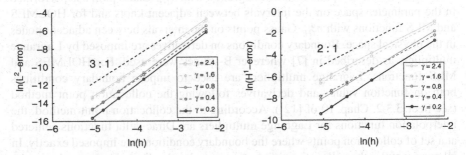

Fig. 3 Comparison of L^2- and H^1-error convergence with HOLMES basis functions for $p = 3$ and varying values of γ

Approximate solutions to (17) can be obtained by a Bubnov-Galerkin method using different types of basis functions. In Fig. 2 we compare the convergence resulting from second-order HOLMES, B-spline, MLS and FEM, respectively. To this end, we use HOLMES and MLS shape functions with the same range $r_{3,HOLMES} = r_{MLS} = 3h$. The HOLMES shape functions are constructed and evaluated with $p = 3$, $\gamma = 0.8$, and a numerical tolerance $\epsilon = 1e-10$. The accuracy of B-splines, MLS functions, and HOLMES is comparable and markedly better than the one achieved by finite element basis functions. We chose $n_{GP} = 2$ for the FEM and B-spline basis functions, and $n_{GP} = 6$ for the HOLMES and MLS functions, respectively. Thus, HOLMES and MLS require a slightly larger range and more integration points than B-splines for a comparable performance. We mention in passing that the computational cost of the evaluation of HOLMES and MLS basis functions in our MATLAB implementation is similar. Figure 3 shows how the accuracy obtained from HOLMES depends on γ. In all cases, optimal convergence is achieved, as expected from the analysis of Sect. 4. Evidently, the error increases and the range decreases with increasing γ, which suggests that larger supports generally result in increased accuracy. In Table 1, range and Gauss-point number requirements are presented for $p = 3$ with varying γ. The number of Gauss quadrature points is chosen as the minimal number for which no ostensible loss

Table 1 Effective range of HOLMES basis functions and number of employed Gauss quadrature points n_{GP} for different values of γ with $p = 3$

γ	2.4	1.6	0.8	0.4	0.2
$r_{3,HOLMES}$	2.0	2.4	3.0	3.8	4.8
n_{GP}	8	6	6	10	>12
ϵ	1e−10	1e−10	1e−10	1e−10	1e−10

Table 2 Effective range of HOLMES basis functions and number of employed Gauss quadrature points n_{GP} for different values of γ with $p = 3$

γ	2.4	1.6	0.8	0.4	0.2
$r_{3,HOLMES}$	2.0	2.4	3.0	3.8	5.1
n_{GP}	16	8	6	5	12
ϵ	1e−10	1e−10	1e−10	1e−10	1e−12

of optimal convergence is observed for any of the discretization lengths employed. Likewise, the tolerance ϵ is set to the minimal value for which no ostensible loss of convergence is observed. For reasons of brevity, we refer to [4] for a more complete discussion regarding the number of Gauss quadrature points necessary to obtain a desired accuracy.

Table 1 and Fig. 3 suggest that $\gamma \approx 0.8$ is a good trade off: for larger γ, the accuracy steadily decreases and the smaller range is offset by an increase in the number of Gauss points; for smaller γ, both range and number of required Gauss points increase steadily without compensating gains in accuracy.

5.2 Elastic Beam Under Sine Load

A first simple test of the performance of HOLMES in fourth-order problems is furnished by the problem of computing the static deflections of an elastic beam. The static displacements $u(x)$ of a linearly elastic Euler-Bernoulli cantilever beam with unit material and geometry parameters under sine load are governed by the partial differential equation

$$u_{xxxx}(x) = \sin(2\pi x),$$
$$x \in [0; 1], u(0) = u_x(0) = u_{xx}(1) = u_{xxx}(1) = 0.$$
(18)

In Table 2 the different parameter values and settings of the HOLMES basis functions applied to this example are presented, and in Figs. 4 and 5, the L^2- and H^1-error norms of numerical solution of (18) with HOLMES, B-spline and MLS basis functions are compared. The general trends are similar to those observed for the elastic rod in Sect. 5.1. Again, $\gamma = 0.8$ seems to be a good parameter choice for HOLMES. Altogether, the performance of HOLMES is slightly better than that of both B-spline and MLS.

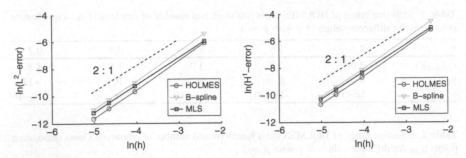

Fig. 4 L^2- and H^1-error convergence of the displacement field of an Euler-Bernoulli cantilever under sine load approximated with different types of basis functions: higher order max-ent functions (HOLMES), B-splines, moving least squares functions (MLS)

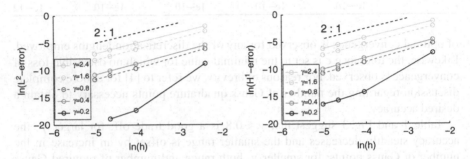

Fig. 5 Comparison of L^2- and H^1-error convergence with HOLMES basis functions for $p = 3$ and varying values of γ

5.3 Membrane Under Sine Load

We extend the analysis of rods presented in Sect. 5.1 to two dimensions by considering the static deformations of a membrane. The example is intended to test the performance of HOLMES in applications involving second-order problems in multiple dimensions. The static displacements $u(x, y)$ of a linearly elastic membrane with unit material and geometry parameters and fixed boundary under sine load in the x and y directions are governed by the partial differential equation

$$u_{xx}(x, y) + u_{yy}(x, y) = \sin(2\pi x) \sin(2\pi y),$$
$$(x, y) \in [0; 1] \times [0; 1], u(0, y) = u(1, y) = u(x, 0) = u(x, 1) = 0. \tag{19}$$

Figure 6 shows convergence plots for HOLMES, B-spline, MLS and FEM obtained with the same parameters and settings as in Sect. 5.1. The general trends follow closely those of the rod problem, which suggests that the observations and conclusions put forth in Sect. 5.1 carry over *mutatis mutandi* to higher dimensions.

We notice that HOLMES, together with MLS, can be applied with unstructured two-dimensional nodal sets, which greatly increases their range of applicability

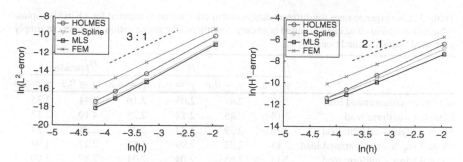

Fig. 6 L^2- and H^1-error convergence of the displacement field of an elastic membrane under sine load approximated with different types of basis functions: higher order max-ent functions (HOLMES), B-splines, moving least squares functions (MLS) and Lagrange polynomials (FEM)

with respect to B-splines. Assessing the convergence properties of HOLMES with non-uniform grids will require more extensive testing and is part of future research. However, one-dimensional patch tests reported in [4] already show that HOLMES can be applied with non-uniform grids.

5.4 Kirchhoff Plate

We proceed to extend the analysis of beams presented in Sect. 5.2 to two dimensions by considering the static deformations of a plate. This example effectively tests the performance of HOLMES in applications involving fourth order problems in multiple dimensions. The static displacements $u(x, y)$ of a linearly elastic Kirchhoff plate with unit material and geometry parameters under a load $q(x, y)$ obey the partial differential equation

$$u_{xxxx}(x, y) + u_{yyyy}(x, y) + 2u_{xxyy}(x, y) = q(x, y)$$
$$(x, y) \in [0; 1] \times [0; 1] + BC \tag{20}$$

where the term BC indicates the boundary conditions on plate displacements and rotations.

The problem of the deflections of a Kirchhoff plate provides a useful test case for investigating issues pertaining to the imposition of boundary conditions. To this end, we solve several examples with the following boundary and loading conditions: clamped boundary with concentrated and uniform load; simply supported boundary on two opposite edges with uniform load; and simply supported boundary on all four edges with concentrated, uniform and sine load. In all calculations, the HOLMES shape functions have been computed using $p = 3$, $\gamma = 0.3$ or $\gamma = 0.8$, and a search radius equal to $5.6h$, which corresponds to a numerical tolerance ϵ of the order of machine precision. Eight different uniform nodal distributions varying from 9×9 to

Table 3 Convergence rates for different displacement error norms computed for a Kirchhoff plate subjected to various loads and different boundary conditions (the abbreviation SS refers to simply supported boundary conditions)

	L^2		H^1		$H^1_{boundary}$	
	$\gamma = 0.3$	$\gamma = 0.8$	$\gamma = 0.3$	$\gamma = 0.8$	$\gamma = 0.3$	$\gamma = 0.8$
Clamped – concentrated load	2.17	2.41	2.03	2.16	3.44	2.35
Clamped – uniform load	2.18	2.45	2.12	2.33	4.10	2.17
SS on 2 edges – uniform load	2.26	2.39	2.23	2.36	2.38	2.34
SS on 4 edges – concentrated load	2.15	2.32	1.99	1.85	2.27	1.94
SS on 4 edges – uniform load	2.13	2.65	2.04	2.31	2.17	2.60
SS on 4 edges – sine load	2.14	2.79	2.11	2.49	2.22	2.97

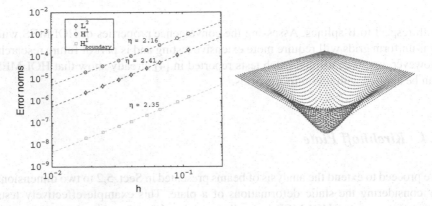

Fig. 7 Elastic Kirchhoff plate clamped on all four edges and subjected to a unit centered concentrated load. Convergence of the L^2- and H^1-error norms of the displacement field (*left*) and deformed configuration for the finest nodal distribution (55 × 55 nodes) used in the convergence analysis (*right*). The order of convergence η for each error norm is also reported

55 × 55 points are used in the convergence analyses. Gauss quadrature is performed with 121 points per element of an auxiliary quadrilateral mesh of the node set. The examples presented in this section have been solved using our in-house C++ code Eureka, which in turn uses the linear solver SuperLU [9].

The convergence of HOLMES applied to the Kirchhoff plate problem is assessed in terms of the convergence of the L^2- and H^1-error norms of the displacement field. The error norms are calculated with respect to the exact analytical solutions for plate bending problems [22]. To make sure that the boundary conditions are effectively enforced, we also compute the H^1-error norm restricted to the relevant boundary of the plate domain. The L^2- and H^1-error norms are computed approximately using 2.5×10^5 uniformly distributed sampling points in the plate domain.

The order of convergence of the error norms computed in the plate examples is reported in Table 3 for two different values of γ. It is noteworthy that the order of convergence is consistently close to or higher than the theoretical convergence order of 2 for both values of γ and throughout the numerical examples, which differ strongly with respect to loading and boundary conditions. As in the case of second-order problems, a comparison of Table 3 and Fig. 4 reveals that the general trends in one dimension carry over to higher dimensions. Figure 7 shows the convergence in the L^2- and H^1-error norms together with the deformed configuration for a clamped plate subjected to a concentrated load. In this case, HOLMES shape functions are computed using $\gamma = 0.8$. The L^2 and H^1 error norms exhibit convergence rates close to or higher than h^2, which is their theoretical convergence rate when quadratic interpolation is applied to fourth order PDEs [20]. The smoothness and regularity of the deformed configuration is also noteworthy.

6 Summary and Conclusions

We have developed a family of approximation schemes, which we have termed High-Order Local Maximum-Entropy Schemes (HOLMES), that deliver approximations of arbitrary order and smoothness, in the sense of convergence in general Sobolev spaces $W^{k,p}$. Thus, HOLMES shape functions span polynomials of arbitrary degree and result in C^k-interpolation of arbitrary degree k of smoothness. HOLMES are constructed in the spirit of the Local Maximum-Entropy (LME) approximation schemes of Arroyo and Ortiz [1] but differ from those schemes in an important respect, namely, the shape functions are not required to be positive. This relaxation effectively bypasses difficulties with lack of feasibility of convex interpolation schemes of high order, albeit at the expense of losing the Kronecker-delta property of LME approximation schemes. Uniform error bounds for the HOLMES approximates and their derivatives up to order k are proven in the extended version of this paper [4]. Moreover, the HOLMES of order k is dense in the Sobolev Space $W^{k,p}$, for any $1 \leq p < \infty$. The good performance of HOLMES relative to other meshfree schemes has also been critically appraised in selected test cases.

On the basis of these results, HOLMES appear attractive for use in a number of areas of application. Thus, because of their meshfree character, HOLMES provide an avenue for highly-accurate simulation of unconstrained flows, be them solid or fluid [14]. In addition, the high-order and degree of differentiability of HOLMES can be exploited in constrained problems, e.g., to account for incompressibility exactly by the introduction of flow potentials, or in higher-order problems such as plates and shells. These potential applications of HOLMES suggest themselves as worthwhile directions for further study.

Acknowledgements The support of the Department of Energy National Nuclear Security Administration under Award Number DE-FC52-08NA28613 through Caltech's ASC/PSAAP Center for the Predictive Modeling and Simulation of High Energy Density Dynamic Response of Materials is gratefully acknowledged. The third author (C.J.C.) gratefully acknowledges the support by the International Graduate School of Science and Engineering of the Technische Universität München.

References

1. M. Arroyo, M. Ortiz, Local *maximum-entropy* approximation schemes: a seamless bridge between finite elements and meshfree methods. Int. J. Numer. Method Eng. **65**(13), 2167–2202 (2006)
2. M. Arroyo, M. Ortiz, Local maximum-entropy approximation schemes. *Meshfree Methods for Partial Differential Equations III*, ed. by M. Griebel, M.A. Schweitzer, vol. 57 (Springer, New York, 2007), pp. 1–16
3. I. Babuška, J.M. Melenk, The partition of unity method. Int. J. Numer. Method Eng. **40**, 727–758 (1997)
4. A. Bompadre, L.E. Perotti, C.J. Cyron, M. Ortiz, Convergent meshfree approximation schemes of arbitrary order and smoothness. Comput. Method Appl. Mech. Eng. **221–222**, 80–103 (2012)
5. A. Bompadre, B. Schmidt, M. Ortiz, Convergence analysis of meshfree approximation schemes. SIAM J. Numer. Anal. 50, 1344–1366 (2012)
6. P.G. Ciarlet, *The Finite Element Method for Elliptic Problems* (SIAM, Philadelphia, 2002)
7. J.A. Cottrell, A. Reali, Y. Bazilevs, T.J.R. Hughes, Isogeometric analysis of structural vibrations. Comput. Method Appl. Mech. Eng. **195**, 5257–5296 (2006)
8. C.J. Cyron, M. Arroyo, M. Ortiz, Smooth, second order, non-negative meshfree approximants selected by maximum entropy. Int. J. Numer. Method Eng. **79**(13), 1605–1632 (2009)
9. J.W. Demmel, S.C. Eisenstat, J.R. Gilbert, X.S. Li, J.W.H. Liu, A supernodal approach to sparse partial pivoting. SIAM J. Matrix Anal. Appl. **20**(3), 720–755 (1999)
10. C.A. Duarte, J.T. Oden, *Hp* clouds – an *hp* meshless method. Numer. Method Partial Differ. Equ. **12**, 673–705 (1996)
11. D. González, E. Cueto, M. Doblaré, A higher order method based on local maximum entropy approximation. Int. J. Numer. Method Eng. **83**(6), 741–764 (2010)
12. A. Huerta, T. Belytschko, S. Fernández-Méndez, T. Rabczuk, *Encyclopedia of Computational Mechanics*, vol. 1, ch. Meshfree methods (Wiley, Chichester, 2004), pp. 279–309
13. E.T. Jaynes, Information theory and statistical mechanics. Phys. Rev. **106**(4), 620–630 (1957)
14. B. Li, F. Habbal, M. Ortiz, Optimal transportation meshfree approximation schemes for fluid and plastic flows. Int. J. Numer. Method. Eng. **83**(12), 1541–1579 (2010)
15. W.K. Liu, S. Li, T. Belytschko, Moving least-square reproducing kernel methods. Part I: methodology and convergence. Comput. Method Appl. Mech. Eng. **143**(1–2), 113–154 (1997)
16. D. Millán, A. Rosolen, M. Arroyo, Thin shell analysis from scattered points with *maximum-entropy* approximants. Int. J. Numer. Method Eng. **85**(6), 723–751 (2011)
17. B. Nayroles, G. Touzot, P. Villon, Generalizing the finite element method: diffuse approximation and diffuse elements. Comput. Mech. **10**(5), 307–318 (1992)
18. V.T. Rajan, Optimality of the Delaunay triangulation in $\mathbb{R}^d$. Discret. Comput. Geom. **12**(2), 189–202 (1994)
19. R.T. Rockafellar, *Convex Analysis* (Princeton University Press, Princeton, 1970)
20. G. Strang, G. Fix, *An Analysis of the Finite Element Method*, (Prentice-Hall, Englewood Cliffs, 1973)
21. N. Sukumar, R. Wright, Overview and construction of meshfree basis functions: from moving least squares to entropy approximants. Int. J. Numer. Method Eng. **70**(2), 181–205 (2007)
22. S. Timoshenko, S. Woinowsky-Krieger, *Theory of Plates and Shells*, 2nd edn. (McGraw-Hill, New York, 1959)
23. H. Wendland, *Scattered Data Approximation* (Cambridge University Press, Cambridge, 2005)

A Meshfree Splitting Method for Soliton Dynamics in Nonlinear Schrödinger Equations

Marco Caliari, Alexander Ostermann, and Stefan Rainer

Abstract A new method for the numerical simulation of the so-called soliton dynamics arising in a nonlinear Schrödinger equation in the semi-classical regime is proposed. For the time discretization a classical fourth-order splitting method is used. For the spatial discretization, however, a meshfree method is employed in contrast to the usual choice of (pseudo) spectral methods. This approach allows one to keep the degrees of freedom almost constant as the semi-classical parameter ε becomes small. This behavior is confirmed by numerical experiments.

Keywords Meshfree discretization • Splitting methods • Nonlinear Schrödinger equations • Soliton dynamics • Semi-classical regime

1 Introduction

In this paper we provide a new numerical method for the solution of the nonlinear Schrödinger equation in the semi-classical regime (see [10])

$$
\begin{cases}
i\varepsilon\partial_t\phi_\varepsilon = -\dfrac{\varepsilon^2}{2}\Delta\phi_\varepsilon + V(x)\phi_\varepsilon - |\phi_\varepsilon|^{2p}\phi_\varepsilon, & x \in \mathbb{R}^d,\ t > 0 \\[2mm]
\phi_\varepsilon(0, x) = \phi_{\varepsilon,0}(x),
\end{cases}
\tag{1}
$$

M. Caliari
Dipartimento di Informatica, Università di Verona, Verona, Italy
e-mail: marco.caliari@univr.it

A. Ostermann · S. Rainer (✉)
Institut für Mathematik, Universität Innsbruck, Innsbruck, Austria
e-mail: alexander.ostermann@uibk.ac.at; stefan.rainer@uibk.ac.at

M. Griebel and M.A. Schweitzer (eds.), *Meshfree Methods for Partial Differential Equations VI*, Lecture Notes in Computational Science and Engineering 89,
DOI 10.1007/978-3-642-32979-1_8, © Springer-Verlag Berlin Heidelberg 2013

where $V(x)$ is a smooth external potential and $0 < p < 2/d$. In particular we are interested in the so-called *soliton dynamics* in which the shape of the time-dependent solution remains close to the initial value. Moreover, such a soliton ϕ is exponentially decaying in space and travels according to Newton's law. We cut off ϕ at some prescribed threshold value (e.g. machine precision) and call the support of this truncated function the *essential support* of ϕ. The essential support is called small, if it is small in comparison to the computational domain, which can be even unbounded. A similar behavior can be observed for solutions of other important classes of partial differential equations, such as wave equations or transport equations. For nonlinear Schrödinger equations, splitting methods (see, e.g. [3, 4, 6, 11, 17]) have shown to be quite efficient and accurate, also in the conservation of the geometric properties such as mass and energy.

For the problem we have in mind, the resolution of the spatial discretization has to be high enough to give a good approximation of the solution whose support is concentrated in a small region. This support depends on the semi-classical parameter ε. The smaller this parameter becomes the smaller the support of the solution gets. On the other hand, in order to keep the degrees of freedom at a reasonable level, it is not possible to have the same high resolution everywhere, in particular, along the trajectory of the soliton dynamics. One should rather consider an adaptive mesh, where most of the discretization points are contained in the moving essential support of the solution. Much more flexible, however, are the so-called meshfree (or meshless) methods, whose advantage is the possibility to easily add or delete discretization points. Nowadays there are many ways to construct meshfree approximations, for instance, by moving least squares or radial basis functions (RBFs). Since the essential support of the solution of problem (1) is compact, we choose compactly supported radial basis functions, developed around 1995 (see [18, 19, 21]).

The overall goal of the proposed approach is to present a method whose degrees of freedom are independent of ε. In addition we introduce a smart procedure to control the spatial accuracy during time evolution, a feature often missing in the method of lines, even if implemented with an adaptive mesh or combined with a meshless approach.

The outline of the paper is as follows. In Sect. 2 we give more details of our equation and state some theoretical facts that we use to solve problem (1). In Sect. 3 we shortly describe the method of RBF discretization. Sections 4 and 5 deal with the numerical solution of problem (1). First, in Sect. 4 we concentrate on the numerical computation of the initial value $\phi_{\varepsilon,0}(x)$. In Sect. 5 we explain how the time evolution is carried out. In particular, we give an introduction to splitting methods and show how the involved semiflows can be computed in the context of meshfree approximations. In the final Sect. 6 we perform some numerical experiments that demonstrate the independence of the degrees of freedom from ε.

2 Soliton Dynamics

We are concerned with the numerical solution of nonlinear Schrödinger equations in the semi-classical limit. The problem has the form (1) with $0 < p < 2/d$. The initial value $\phi_{\varepsilon,0}(x)$ is a bump-like function

$$\phi_{\varepsilon,0}(x) = r\left(\tfrac{x-\bar{x}}{\varepsilon}\right)e^{\frac{i}{\varepsilon}x\cdot\bar{v}}, \quad \bar{x}, \bar{v} \in \mathbb{R}^d. \tag{2}$$

Here, $r(x)$ denotes the radial part of the ground state solution of

$$\begin{cases} i\partial_t\phi = -\dfrac{1}{2}\Delta\phi - |\phi|^{2p}\phi \\ \|\phi\|_{L^2}^2 = m, \qquad m > 0 \text{ given.} \end{cases} \tag{3}$$

The ground state has the form $\phi(t,x) = u(x)e^{-i\nu t}$ and minimizes the energy functional

$$E(\phi) = E(u) = \frac{1}{2}\int_{\mathbb{R}^d}|\nabla u|^2\mathrm{d}x - \frac{1}{p+1}\int_{\mathbb{R}^d}|u|^{2p+2}\mathrm{d}x.$$

In this setting the following assertion holds.

Proposition 1 (see [8, Theorem 1]). *Let $\phi_{\varepsilon}(t,x)$ be the solution of* (1) *with initial value* (2). *Then, there exists a family of shifts $\theta_{\varepsilon}:\mathbb{R}^+ \to [0, 2\pi)$ such that, as ε tends to zero,*

$$\left\|\phi_{\varepsilon}(t,x) - r\left(\tfrac{x-x(t)}{\varepsilon}\right)\exp\left(\tfrac{i}{\varepsilon}(x\cdot\dot{x}(t) + \theta_{\varepsilon}(t))\right)\right\|_{\mathbb{H}_{\varepsilon}} = \mathcal{O}(\varepsilon),$$

locally uniformly in the variable t.

Here,

$$\|\phi\|_{\mathbb{H}_{\varepsilon}}^2 = \varepsilon^{2-d}\|\nabla\phi\|_{L^2}^2 + \varepsilon^{-d}\|\phi\|_{L^2}^2,$$

and $x(t)$ is the solution of Newton's law

$$\begin{cases} \ddot{x}(t) = -\nabla V(x(t)), \\ x(0) = \bar{x}, \\ \dot{x}(0) = \bar{v}. \end{cases} \tag{4}$$

This dynamical behavior, where the shape of $|\phi_{\varepsilon}(t,x)|^2$ remains close to $|\phi_{\varepsilon,0}(x)|^2$, is typically known as *soliton dynamics*. The case where the initial value is a multibump, say

$$\phi_{\varepsilon,0}(x) = \sum_{j=1}^{\ell} r_j \left(\frac{x - \bar{x}_j}{\varepsilon} \right) e^{\frac{1}{\varepsilon} x \cdot \bar{v}_j}, \tag{5}$$

where $r_j(x)$ minimizes $E(u)$ under the restriction $\|u\|_{L^2}^2 = m_j$, is studied, e.g., in [10, 14].

3 Meshfree Approximation

For the numerical representation of a function $f : \mathbb{R}^d \to \mathbb{R}$ we use the concept of RBF interpolation. Given a set of interpolation points $\Xi = \{\xi_1, \ldots, \xi_n\}$ and a radial function

$$\Psi^\alpha(x) = \psi\left(\alpha \|x\|\right), \qquad \alpha > 0$$

we construct an interpolant

$$p(x) = \sum_{\xi \in \Xi} \lambda_\xi \Psi_\xi^\alpha(x), \qquad \Psi_\xi^\alpha(x) = \Psi^\alpha(x - \xi)$$

satisfying

$$p(\xi_i) = f(\xi_i), \qquad i = 1, \ldots, n.$$

For more details on RBF interpolation, in particular for the computation of the coefficients $[\lambda_\xi]_{\xi \in \Xi}$, we refer to [9, 15, 20].

In order to save storage one aims to minimize the required number of interpolation points under the condition $|p(x) - f(x)| \leq$ tol for a given tolerance tol > 0 and all $x \in \mathbb{R}^d$. This can be done using a residual subsampling approach as described in [13]. For the convenience of the reader, a summary of this approach is given in Appendix A. Moreover with the help of the shape parameter α we can adapt the form of the basis function in such a way that the error is minimized for a fixed set of interpolation points.

For the practical implementation of the method we take the compactly supported RBF

$$\psi(r) = (1 - r)_+^6 (35r^2 + 18r + 3),$$

also known as Wendland's function $\varphi_{3,2}$. Here x_+^p denotes the truncated power function, i.e.

$$x_+^p = \begin{cases} x^p & \text{if } x > 0, \\ 0 & \text{else.} \end{cases}$$

For more details on compactly supported RBFs we refer to [9, 15, 19, 20].

4   Approximation of the Initial Value

The considered initial value (2) is the product of the ground state of (3) at time $t = 0$ and a phase factor. This ground state, however, does not depend on ε. Hence we can use standard methods to compute it.

4.1   Approximation of the Ground State Solution

There are several approaches to compute the ground state solution of (3). One possibility consists in directly minimizing the Euler–Lagrange function

$$\mathscr{E}(u, \mu) = E(u) + \mu(\|u\|_{L^2}^2 - m)$$

by a Newton-like method with approximate line-search strategy, as done, for instance, in [12]. Here, we use an approach similar to the imaginary time method (see [2]). It is described in [10] and relies on the solution of the parabolic PDE

$$
\begin{cases}
\partial_\tau u = \dfrac{1}{2}\Delta u + u^{2p+1} + \mu(u)u, & \tau > 0 \\[2mm]
u(0, x) = u_0(x), & \|u_0\|_{L^2}^2 = m
\end{cases}
\tag{6}
$$

with

$$\mu(u) = \frac{\frac{1}{2}\int_{\mathbb{R}^d}|\nabla u|^2 \mathrm{d}x - \int_{\mathbb{R}^d}|u|^{2p+2}\mathrm{d}x}{\|u\|_{L^2}^2}.$$

It was be shown in [10] that $u(\tau, x)$ has constant mass m and its energy decreases in time. Moreover, the steady-state solution $r(x) = \lim_{\tau \to \infty} u(\tau, x)$ together with $\kappa = \mu(r(x))$ gives the desired ground state solution $r(x)e^{-i\kappa t}$ of (3), see [10].

In both cases, a spatial discretization of the unknown function is necessary. For this, spectral methods, based on Hermite or Fourier decompositions are quite standard. Hermite functions are to be preferred when an external harmonic (or quasi-harmonic) potential, that is a term of the form $H(x)\phi$ with $H(x) \approx \frac{|x|^2}{2}$ is present in (3); see, for instance, [12]. In our case of the free-potential equation the use of Fourier modes is suggested. In fact, the ground state solution of (3) exponentially decays to zero and is symmetric (see [10]). This means that the computational domain can be safely truncated to a reasonable size, corresponding to a modest number of Fourier modes to be used. We further note that the computational domain required at this stage needs just to contain the essential support of the ground state solution and not the whole trajectory of the soliton dynamics.

After the spatial discretization of (6) we end up with a large system of ODEs of the form

$$\begin{cases} w'(t) = Dw(t) + g(w(t)), & t > 0 \\ w(0) = w_0 \end{cases}$$

whose solution can be approximated with an exponential integrator (see Appendix B). Here $w(t)$ denotes the vector of Fourier coefficients of $u(t, x)$. The diagonal matrix D contains the eigenvalues of the Laplace operator (divided by 2) with respect to the Fourier eigenfunctions and $g(w(t))$ is the truncated discrete Fourier expansion of the nonlinear part of (6). Exponential integrators (see the survey paper [16]) are particularly suited in this situation, because they are explicit and can manage stiff problems without any restriction on the time step size. Moreover, in the case of a diagonal matrix, the computation of matrix exponential-like functions is straightforward. We use the exponential Runge–Kutta method of order 2 as described in Appendix B and the embedded exponential Euler method as error estimate (see [16]). This is particularly useful when computing a steady-state solution. In this situation the time step sizes can be chosen larger and larger when approaching the steady-state. We stop the integration at time $t = \bar{t}$ and take $w(\bar{t})$ as "steady-state" solution if the condition

$$\|Dw(\bar{t}) + g(w(\bar{t}))\| \leq \delta$$

is satisfied for a given tolerance δ. The initial value $u(0, x)$ in (6) can be arbitrary, but it appears reasonable to take a solution with small energy. If we restrict ourselves to the Gaussian family of mass m

$$G_\sigma(x) = \frac{\sqrt{m\sigma^d}}{\sqrt[4]{\pi^d}} e^{-\sigma^2 |x|^2/2}, \quad \sigma > 0$$

it is possible to analytically compute the energy $E(G_\sigma)$ and to minimize it, for $d = 1, 2, 3$ (see [10]). The exponential decay σ for which the energy of the corresponding Gaussian function is minimal gives also an indication of the size of the required computational domain.

4.2 Choice of the Interpolation Points and the Shape Parameter

From the above construction, we have the radial part r of the ground state available on a fine grid $\mathscr{G}$. Next we replace r by its RBF interpolant p on an reasonable small set of interpolation points Ξ. To construct this set, we use a multilevel iteration (see [15]). We select successively interpolation points from $\mathscr{G}$ and add them to Ξ until the accuracy requirement

$$|p(x) - \phi_{\varepsilon,0}(x)| \leq \text{tol}$$

is satisfied for all $x \in \mathcal{G}$. In order to be able to use the resulting set Ξ for the solution of (1), we have to rescale the spatial variable. For $y := \frac{x}{\varepsilon}$, we get

$$\chi(y) = \phi_{\varepsilon,0}(x) = r(y - \bar{y})e^{i\bar{y}\bar{v}}$$

which can be represented on the set of interpolation points $\Xi_\varepsilon := \varepsilon \Xi$. Note that in order to keep the interpolation error small, the radial functions have to be rescaled as well. For arbitrary radial functions this can be done by adjusting the shape parameter α, using a standard one-dimensional minimization procedure. We take the golden section search method starting from an admissible interval $[0, \alpha^*]$, where α^* is chosen inverse proportional to ε.

5   Approximation of the Time Dependent Solution

For the numerical solution of (1), splitting methods are well established (see, e.g., [17, Chap. 3]). In this section we give a short introduction to splitting methods, and we describe how the involved semiflows can be computed in the context of meshfree approximations. Based on this we explain how the proposed numerical method is constructed.

5.1   Splitting Methods

The idea behind splitting is to decompose the right-hand side of a differential equation into two parts and to split the problem into two subproblems. The proper combination of their solutions gives an approximation to the solution of the original problem of desired order.

In our case we split equation (1) into

$$i\varepsilon \partial_t \phi_\varepsilon^{[1]}(t, x) = -\frac{\varepsilon^2}{2} \Delta \phi_\varepsilon^{[1]}(t, x) \tag{7a}$$

and

$$i\varepsilon \partial_t \phi_\varepsilon^{[2]}(t, x) = V(x)\phi_\varepsilon^{[2]}(t, x) - |\phi_\varepsilon^{[2]}(t, x)|^{2p}\phi_\varepsilon^{[2]}(t, x). \tag{7b}$$

The linear problem (7a) is usually solved by (pseudo) spectral methods, relying on Hermite or Fourier basis functions (see, e.g., [11, 17]). Although the mass of the solution is mainly concentrated in a small (time dependent) region, the computational domain has to be large enough to cover at least a neighborhood of the whole trajectory $x(t)$. This might lead to an unreasonable number of spectral modes, if one requires the solution to have the same accuracy as the initial value.

Table 1 Coefficients of SRKN$_6^b$, see [7]

i	a_i	b_i
1	0.245298957184271	0.0829844064174052
2	0.604872665711080	0.396309801498368
3	$0.5 - (a_1 + a_2)$	-0.0390563049223486
4	a_3	$1 - 2(b_1 + b_2 + b_3)$
5	a_2	b_3
6	a_1	b_2
7	0	b_1

In order to keep the degrees of freedom small, we use a meshfree approximation based on compactly supported radial basis functions instead.

For a real potential V, the modulus of the solution of (7b), $|\phi_\varepsilon^{[2]}(t, x)|$, stays constant in time, pointwise in x. Thus problem (7b) is a linear differential equation with time independent coefficients depending on a parameter $x \in \mathbb{R}^d$, and its solution can be computed exactly.

For the numerical approximation of $\phi_\varepsilon(t_{n+1}, x)$ we have to combine the exact flows of the split problem, namely

$$\partial_t \phi_\varepsilon^{[1]} = \frac{i\varepsilon}{2} \Delta \phi_\varepsilon^{[1]} = L\phi_\varepsilon^{[1]}, \qquad \phi_\varepsilon^{[1]}(t_n, x) = w_1(x), \qquad (8a)$$

$$\partial_t \phi_\varepsilon^{[2]} = B\phi_\varepsilon^{[2]}, \qquad \phi_\varepsilon^{[2]}(t_n, x) = w_2(x) \qquad (8b)$$

with the multiplication operator $B(x) = V(x) - |w_2(x)|^{2p}$ and appropriately chosen initial values w_1 and w_2. As time integration we use the fourth-order symmetric Runge–Kutta–Nyström method SRKN$_6^b$ which has the form

$$\left(\prod_{i=1}^{7} e^{a_i k L} e^{b_i k B} \right) \phi_\varepsilon(t_n, x).$$

Here $k = t_{n+1} - t_n$ denotes the time step size. The coefficients a_i, b_i are given in Table 1. For more details about the method we refer to [7].

5.2 Computation of the Semiflows

For the spatial discretization of (8) we use radial basis functions centered at the set of interpolation points Ξ_ε. This allows us to identify a vector $[v_\xi]_{\xi \in \Xi}$ with its interpolant $v(\cdot)$ and vice versa.

Since the multiplication operator B in (8b) acts pointwise, the numerical computation of $\phi_\varepsilon^{[2]}(t_n + b_i k, \xi) = e^{b_i k B(\xi)} w_2(\xi)$ is trivial for every $\xi \in \Xi_\varepsilon$.

To approximate the solution of (8a)

$$\phi_\varepsilon^{[1]}(t_n + a_i k, \xi) = \left(e^{a_i k L} w_1\right)(\xi), \qquad \xi \in \varXi_\varepsilon$$

we use polynomial interpolation at conjugate pairs of Leja points (see [5,13]). Just as direct Krylov methods (see [16] and references therein) and truncated Taylor series approximations (see [1]), Leja interpolation is based on successive applications of L to an interpolant of a given vector $[v_\xi]_{\xi \in \varXi_\varepsilon}$. In case of meshfree methods this can be realized as follows: we construct an interpolant

$$p(x) = \sum_{\xi \in \varXi_\varepsilon} \lambda_\xi \varPsi_\xi^\alpha(x), \qquad p(\xi) = v_\xi, \quad \xi \in \varXi_\varepsilon \tag{9}$$

on the given set of interpolation points $\varXi_\varepsilon$. The coefficients $[\lambda_\xi]_{\xi \in \varXi_\varepsilon}$ are computed from the collocation conditions. With A denoting the interpolation matrix $\varPsi_\xi^\alpha(\eta)$, $\xi, \eta \in \varXi_\varepsilon$, we approximate $\frac{i\varepsilon}{2} \Delta v(x)$ by

$$Lp(x) = \frac{i\varepsilon}{2} \Delta p(x) = \sum_{\xi \in \varXi_\varepsilon} \left[A^{-1} v\right]_\xi \left(\frac{i\varepsilon}{2} \Delta \varPsi_\xi^\alpha\right)(x),$$

hence

$$Lp(\cdot)|_{\xi \in \varXi_\varepsilon} = A_L A^{-1} p(\cdot)|_{\xi \in \varXi_\varepsilon} = A_L A^{-1} v$$

with the matrix $A_L = \frac{i\varepsilon}{2} \Delta \varPsi_\xi(\eta)$, $\xi, \eta \in \varXi_\varepsilon$.

5.3   Time Evolution of the Solution

The essential support of the solution of (1) for a fixed time becomes smaller with decreasing ε. The trajectory, however, stays close to the solution of (4) on compact time intervals. Since the modulus of the solution does not change much during time evolution we want to use a similar number of interpolation points for any ε and arbitrary time intervals. This will be done in the following way. We start with the set of interpolation points for the ground state which is independent of ε. This set is scaled according to the procedure described in Sect. 4.2. This gives an appropriate set of interpolation points for the initial value $\phi_{\varepsilon,0}(x)$. Next we combine the splitting method described before and the concept of a meshfree integrator as in [13] to advance the solution in time.

Starting from an approximation at time t_n we perform a time step and get a set of interpolation points for the current solution. Depending on the step size this set of interpolation points might be too large to represent the actual solution (in order to perform a time step the set of interpolation points has to cover the essential support of the solution for the whole step). To reduce this set we use the concept of residual subsampling to find a reliable set of interpolation points for the actual solution. As described in Appendix A, the subsampling procedure requires a set of candidate interpolation points. This set is obtained by propagating the set of interpolation points of the previous time according to (4). More precisely, we compute the solution of (4) for the actual step and add the difference between the actual solution and the solution for the previous time to all the interpolation points. The shifted set forms our set of candidate interpolation points.

After this subsampling step, we have found an approximation at time t_{n+1}, represented by an appropriate number of radial basis functions. This procedure is repeated until we reach the final time.

6 Numerical Experiments

In this section we give some numerical experiments that demonstrate the power of the method. Throughout this section we consider equation (1) in two dimensions with $p = 0.2$ and the harmonic potential

$$V(x, y) = \frac{(1.5x)^2 + y^2}{2}.$$

We integrate from $t = 0$ to 2π. The initial value is the bump-like function (2) with $m = 1$, $\bar{x} = [-2.5, -2.5]^T$ and $\bar{v} = [0, 0]^T$. The trajectory for this potential is a periodic Lissajous curve, starting at $[-2.5, -2.5]^T$ and ending at $[2.5, -2.5]^T$. In Fig. 1 the time evolutions for $\varepsilon = 10^{-2}$ and $\varepsilon = 10^{-3}$ are shown. The spatial tolerance is $2.5 \cdot 10^{-4}$ and the time step size is $k = 10^{-3}$.

Note that for moderate values like $\varepsilon = 10^{-2}$, standard grid-based methods have no problems, too. However, for decreasing values of ε, the computational cost increases considerably. In fact, the trajectories lie in the same computational domain, but the essential support of the solution becomes smaller. Therefore, the number of grid points, or, equivalently, the number of spectral modes, has to be increased in order to have the same spatial resolution during time evolution. The growth of the degrees of freedom can be dramatic, especially in the three dimensional case.

On the other hand, for the proposed method the number of interpolation points is in practice independent of ε, as can be seen in Fig. 2, on the left. For the same problem as described above and a range of ε from 2^{-6} to 2^{-16}, we state the number of interpolation points needed to meet the prescribed spatial tolerance tol.

In the right plot in Fig. 2, we see that the shape parameter α of the compactly supported radial basis functions $\Psi^\alpha(x)$ scales as ε^{-1}, as expected from Sect. 4.2.

Still an open problem is the time step size. The error in time is inverse proportional to ε. This means that a small ε will lead to a large number of time steps. This requires a new time integrator which is work in progress.

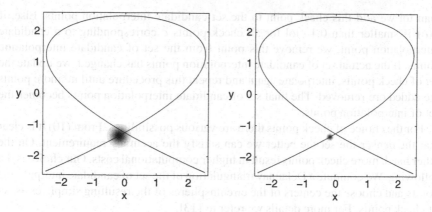

Fig. 1 Solutions of (1) at $t = 1$ with the whole trajectory of (4) for $\varepsilon = 10^{-2}$ (*left*) and $\varepsilon = 10^{-3}$ (*right*) for spatial tolerance tol $= 2.5 \cdot 10^{-4}$ and time step size $k = 10^{-3}$

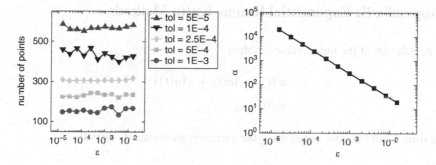

Fig. 2 Number of required interpolation points for the representation of the initial value $\phi_{\varepsilon,0}(x)$, for $\varepsilon = 2^{-j}$, $j = 6, \ldots, 16$ and different prescribed spatial tolerances tol (*left*); value of optimal shape parameter α of the basis functions for the representation of the initial value $\phi_{\varepsilon,0}(x)$, for different prescribed values of $\varepsilon = 2^{-j}$, $j = 6, \ldots, 16$ (*right*)

Appendix A: Residual Subsampling

The idea of residual subsampling is the following. Given a set of candidate interpolation points and a function $f : \mathbb{R}^d \to \mathbb{R}$, find a reasonable set of interpolation points in order that the resulting interpolant p fulfills a certain accuracy requirement. Our aim is to construct an interpolant that satisfies

$$\theta(x) = |f(x) - p(x)| \leq \text{tol}, \qquad x \in \mathbb{R}^d. \tag{10}$$

For this purpose we consider in addition a set of check points $\mathscr{C}$ which is used to check (10). Every candidate interpolation point has at least one check point. Using the set of candidate interpolation points we construct an interpolant of f and evaluate it at the set of check points. If the value $\theta(c)$ for some $c \in \mathscr{C}$ is larger than tol we add this check point to the set candidate interpolation points. Else, if $\theta(c)$ is smaller than $0.1 \cdot$ tol for all check points c corresponding to a candidate interpolation point, we remove this point from the set of candidate interpolation points. If the actual set of candidate interpolation points has changed, we update the set of check points, interpolate again and repeat this procedure until no more points are added ore removed. The final set of candidate interpolation points becomes the set of interpolation points.

For the choice of check points there are various possibilities. From (10) it is clear that the denser the set the better we can satisfy the accuracy requirement. On the other hand, more check points result in higher computational costs. Our choice is the following. We compute a Delaunay triangulation of the set of candidate interpolation points and choose the centers of the circumspheres of the resulting simplices as set of check points. For more details we refer to [13].

Appendix B: Exponential Runge–Kutta Methods

The solution of the initial value problem

$$u'(t) = Au(t) + g(u(t))$$
$$u(0) = u_0$$

at time h can be represented using the variation-of-constants formula

$$u(h) = e^{hA}u_0 + \int_0^h e^{(h-\tau)A}g(u(\tau))d\tau.$$

If we replace $u(\tau)$ by the known value u_0 we get an approximation of order 1

$$u_h = e^{hA}u_0 + \int_0^h e^{(h-\tau)A}g(u_0)d\tau = e^{hA}u_0 + h\varphi_1(hA)g(u_0),$$

also called the *exponential Euler method*. Here $\varphi_1(z)$ denotes the entire function

$$\varphi_1(z) = \frac{e^z - 1}{z}.$$

An exponential Runge–Kutta method of order 2 is given by

$$u_h = e^{hA}u_0 + h(\varphi_1(hA) - \varphi_2(hA))g(u_0) + h\varphi_2(hA)g(U)$$
$$U = e^{hA}u_0 + h\varphi_1(hA)g(u_0)$$

with

$$\varphi_2(z) = \frac{\varphi_1(z) - 1}{z}.$$

For more details we refer to [16]. In order to compute $e^{hA}w$, $\varphi_1(hA)w$ and $\varphi_2(hA)w$ for a given vector w, we use the *real Leja point method* which is based on Newton's interpolation formula at a sequence of Leja points, see [5, 13].

Acknowledgements The work of Stefan Rainer was partially supported by the Tiroler Wissenschaftsfond grant UNI-0404/880.

References

1. A.H. Al-Mohy, N.J. Higham, Computing the action of the matrix exponential, with an application to exponential integrators. SIAM J. Sci. Comput. **33**, 488–511 (2011)
2. W. Bao, Q. Du, Computing the ground state solution of Bose–Einstein condensates by a normalized gradient flow. SIAM J. Sci. Comput. **25** 1674–1697 (2004)
3. W. Bao, J. Shen, A fourth-order time-splitting Laguerre–Hermite pseudospectral method for Bose–Einstein condensates. SIAM J. Sci. Comput. **26** 2010–2028 (2005)
4. W. Bao, D. Jaksch, P. Markowich, Numerical solution of the Gross–Pitaevskii equation for Bose–Einstein condensation. J. Comp. Phys. **187** 318–342 (2003)
5. L. Bergamaschi, M. Caliari, M. Vianello, Interpolating discrete advection-diffusion propagators at Leja sequences. J. Comput. Appl. Math. **172**, 79–99 (2004)
6. C. Besse, B. Bidégaray, S. Descombes, Order estimates in time of splitting methods for the nonlinear Schrödinger equation. SIAM J. Numer. Anal. **40**, 26–40 (2002)
7. S. Blanes, P.C. Moan, Practical symplectic partitioned Runge–Kutta and Runge–Kutta–Nyström methods. J. Comput. Appl. Math. **142** 313–330 (2002)
8. J. Bronski, R. Jerrard, Soliton dynamics in a potential. Math. Res. Lett. **7**, 329–342 (2000)
9. M.D. Buhmann, *Radial Basis Functions: Theory and Implementations* (Cambridge University Press, Cambridge, 2003)
10. M. Caliari, M. Squassina, Numerical computation of soliton dynamics for NLS equations in a driving potential. Electron. J. Differ. Equ. **89**, 1–12 (2010)
11. M. Caliari, C. Neuhauser, M. Thalhammer, High-order time-splitting Hermite and Fourier spectral methods for the Gross–Pitaevskii equation. J. Comput. Phys. **228**, 822–832 (2009)
12. M. Caliari, A. Ostermann, S. Rainer, M. Thalhammer, A minimisation approach for computing the ground state of Gross–Pitaevskii systems. J. Comput. Phys. **228**, pp. 349–360 (2009)
13. M. Caliari, A. Ostermann, S. Rainer, *Meshfree Exponential Integrators*, to appear in SIAM J. Sci. Comput. (2011)
14. M. Caliari, A. Ostermann, S. Rainer, Meshfree integrators. Oberwolfach Rep. **8**, 883–885 (2011)
15. G.E. Fasshauer, *Meshfree Approximation Methods with MATLAB* (World Scientific, Hackensack, 2007)
16. M. Hochbruck, A. Ostermann, Exponential integrators. Acta Numer. **19**, 209–286 (2010)
17. C. Lubich, *From Quantum to Classical Molecular Dynamics: Reduced Models and Numerical Analysis* (European Mathematical Society, Zürich, 2008)
18. R. Schaback, Creating surfaces from scattered data using radial basis functions, in *Mathematical Methods for Curves and Surfaces*, ed. by M. Dæhlen, T. Lyche, L.L. Schumaker (Vanderbilt University Press, Nashville, 1995), pp. 477–496
19. H. Wendland, Piecewise polynomial, positive definite and compactly supported radial functions of minimal degree. Adv. Comput. Math. **4**, 389–396 (1995)
20. H. Wendland, *Scattered Data Approximation* (Cambridge University Press, Cambridge, 2005)
21. Z. Wu, Compactly supported positive definite radial functions. Adv. Comput. Math. **4** 283–292 (1995)

A Meshless Discretization Method for Markov State Models Applied to Explicit Water Peptide Folding Simulations

Konstantin Fackeldey, Alexander Bujotzek, and Marcus Weber

Abstract Markov State Models (MSMs) are widely used to represent molecular conformational changes as jump-like transitions between subsets of the conformational state space. However, the simulation of peptide folding in explicit water is usually said to be unsuitable for the MSM framework. In this article, we summarize the theoretical background of MSMs and indicate that explicit water simulations do not contradict these principles. The algorithmic framework of a meshless conformational space discretization is applied to an explicit water system and the sampling results are compared to a long-term molecular dynamics trajectory. The meshless discretization approach is based on spectral clustering of stochastic matrices (MSMs) and allows for a parallelization of MD simulations. In our example of Trialanine we were able to compute the same distribution of a long term simulation in less computing time.

Keywords Molecular dynamics • Conformation dynamics • Robust perron cluster analysis • Markovianity • Transfer operator • Partition of unity

1 Introduction

The dynamics of biomolecules is inherently multiscale since it involves time scales ranging from femtoseconds (bond-bond oscillations) to microseconds (protein folding, binding process). In the literature many methods can be found, which seek to capture this multiscale behavior. In molecular dynamics (MD) [12, 20] for instance, the integration step for the simulation is bounded to femtoseconds, which complicates the simulation of the above mentioned folding (e.g. [34]).

K. Fackeldey (✉) · A. Bujotzek · M. Weber
Zuse Institute Berlin (ZIB), Takustraße 7, 14195 Berlin, Germany
e-mail: fackeldey@zib.de

M. Griebel and M.A. Schweitzer (eds.), *Meshfree Methods for Partial Differential Equations VI*, Lecture Notes in Computational Science and Engineering 89,
DOI 10.1007/978-3-642-32979-1_9, © Springer-Verlag Berlin Heidelberg 2013

On the other hand, coarse graining models [6, 11], which allow for larger or even unphysical time steps, suffer from the fact that the description may be too coarse and thus information is lost. In the last few years, much effort has been invested into combining the advantages of the coarse graining models with the ones of molecular dynamics. One approach is based on so-called Markov State Models (MSM) e.g. [24]. In particular, we mention the works in the context of Folding at Home [1, 3], the works in the context of conformation dynamics [8, 28, 29] or approaches by other groups also using MSM e.g. [23, 31]. We remark that this list is far from being exhausive. In such models, the different conformational states of a molecular system and its transition probabilities between the states are incorporated, so that the "future" of the system over an arbitrary time span can be predicted. Thereby, it is assumed, that the system has a Markov property, which means that the probability of the system to switch to the next state depends on the current state only. In this context a state (or conformation) is an almost invariant subset in the phase space, where the system is not stable but almost stable (metastable). The performance of this method has been shown in various examples for molecular systems in vacuum. However, to find a MSM which represents the dynamics of a molecule simulated in (explicit) solvent is more difficult [7]. Explicit solvent can spoil the Markov property, such that the long-term behavior of the system is not reproduced correctly by the MSM [27, 38]. A classical MSM is a projection of the continuous dynamics to a low number of discrete (Markov) states, i.e. to a low number of subsets of the conformational space. In this article we want to fill this gap and present simulations of Trialanine in water. In the classical theory of Markov State Models [7, 33], the clusters, corresponding to the molecular conformations of the system, are represented by a collection of characteristic basis functions that yield 1 if the state belongs to the one conformation and zero otherwise. Here, we relax the condition, by softening the hard clustering [37]. More precisely, we allow a state to belong to more than one conformation and assign a degree of membership to each state. As a consequence, this soft clustering allows for a faithful representation of intermediate states, i.e. states which lie in transition regions between metastable conformations. These intermediate states are prevalent in simulation of biomolecules in water, where hydrogen bonds of the surrounding liquid can influence the stability of the molecule [26]. The existence of intermediate states allows for a correct representation of the long-term behavior of molecular systems including explicit solvent. This will be exemplified with Trialanine in water.

2 Basics of Conformation Dynamics

Let us assume, that the dynamics of a molecular system is given by the Hamiltonian function:

$$H(q, p) = V(q) + K(p) \tag{1}$$

where $K(p)$ is the kinetic energy, which depends on the N generalized momenta $p = (p_1, \ldots, p_N) \in \mathbb{R}^{3N}$ and $V(q)$ is the potential energy depending on the N

generalized positions $q = (q_1, \ldots, q_N) \in \mathbb{R}^{3N}$. The term $H(q, p)$ is denoted as the internal energy of the system $x = (q, p) \in \Gamma = \Omega \times \mathbb{R}^{3N}$, where Γ is the state space and $\Omega \subset \mathbb{R}^{3N}$ is the position space. We consider a canonical ensemble, where the number of particles, the volume and the temperature is kept constant. According to the Boltzmann distribution, the positions q and momenta p of each atom in a molecule are then given by:

$$\mu(x) = \mu(q, p) \propto \exp(-\beta H(q, p)). \tag{2}$$

Here, $\beta = 1/k_B T$ is the inverse of temperature T times Boltzmann constant k_B. This canonical density can be split into a distribution of positions $\pi(q)$ and momenta $\eta(p)$, where

$$\pi(q) \propto \exp(-\beta V(q)) \text{ and } \eta(p) \propto \exp(-\beta K(p)). \tag{3}$$

We can now introduce the Hamiltonian equations of motions by

$$\dot{q} = \frac{\partial H}{\partial p} \quad , \quad \dot{p} = -\frac{\partial H}{\partial q}. \tag{4}$$

For an initial state $x(0) = x_0$, the Hamiltonian flow ϕ^t over time span t is given by

$$x(t) = (q(t), p(t)) = \phi^t x_0,$$

where ϕ^t conserves the energy and is reversible. Moreover, the Hamiltonian flow conserves the energy, which implies that the Lebesgue measure on Γ is invariant under ϕ^t.

Let Π_q be the projection of the state (q, p) onto the position q, let Φ^τ be the discrete Hamiltonian flow for a time span τ. It is realized by a velocity verlet integrator with a constant time step and let further p be chosen randomly according to the distribution $\eta(p)$, then

$$q_{i+1} = \Pi_q \Phi^\tau(q_i, p_i)$$

describes a Markov process where the $i + 1$th state depends on the preceding ith state only. Since $\mu(q, p)$ is symmetric in the sense $\mu(q, p) = \mu(q, -p)$ it can be shown, that the reduced density

$$\pi(q) = \int_{\mathbb{R}^{dN}} \mu(p, q) \, dp$$

is smooth, positive, finite and $\pi(q) = 1$, which allows us to define the transition operator [28]

$$T^\tau u(q) = \int_{\mathbb{R}^{dN}} u(\Pi \Phi^{-\tau}(q, p)) \eta(p) dp, \tag{5}$$

where $u(q)$ is a function on Ω. The operator in (5) can be explained as follows: We apply the Hamiltonian dynamics backwards to a lag time τ and obtain $\Phi^{-\tau}(q, p)$, which is then projected by Π_q onto the state space. The integral then, averages over all possible initial momentum variables with given Boltzmann distribution η.

This operator on the weighted Hilbert space $L^2_\pi(\Omega)$, which is equipped with the inner product

$$\langle u, v \rangle_\pi := \int_\Omega u(q)v(q)\pi(q)dq \,,$$

is bounded, linear and self-adjoint. This allows us to compute the transition probability between two sets A and B by

$$p(A, B, t) = \frac{\int_A \chi_B(\Phi^\tau(x))\eta(p)dp}{\int_A \eta(p)dp} = \frac{\langle T\chi_A, \chi_B \rangle_\pi}{\langle \chi_A, \chi_B \rangle_\pi},$$

where χ_A is the characteristic function being 1 in A and 0 otherwise. Based on this, we define a subset $B \subset \Omega$ as almost invariant, if

$$p(B, B, t) \approx 1.$$

and a metastable conformation as a function $C : \Omega \rightarrow [0, 1]$ which is nearly invariant under the transfer operator T^τ, i.e.

$$T^\tau C(q) \approx C(q). \tag{6}$$

The fundamental idea behind this formulation is, that the transfer operator T^τ in (5) is a *linear* operator, although the ordinary differential equation (4) is (extremely) *non-linear*.

2.1 Discretization of the State Space

The linearization by the above introduced transfer operator allows for a Galerkin discretization of T^τ and thus for a numerical approximation of eigenfunctions and eigenvalues of the discrete spectrum of T^τ.

We thus decompose the state space into metastable sets. More precisely, with respect to (6), we aim at a set $\{C_1, \ldots, C_{n_c}\}$ such that they form a partition of unity, i.e.

$$\sum_{J=1}^{n_c} C_J(q) = 1_\Omega \quad \forall q \in \Omega \,,$$

where $C_J : \Omega \rightarrow [0, 1]$ is a function and $C_J(q) \geq 0 \quad \forall q \in \Omega$.

Of course, this decomposition is not given in advance, i.e. the metastable sets have to be detected. To do so, we discretize the state space by meshfree basis functions.

Let $\{x_i\}_{i=1}^N$ be a set of points from which we construct the meshfree basis functions by Shepard's approach [30] :

$$\varphi_i(q) = \frac{\exp(-\alpha\|q - x_i\|)}{\sum_{i=1}^N \exp(-\alpha\|q - x_i\|)}$$

such that

$$\forall q \in \Omega : \quad \sum_{i=1}^N \varphi_i(q) = 1 \text{ and } \bigcup_{i=1}^N \text{supp}(\varphi_i) = \Omega .$$

Finally, this allows us to represent each C_J as a linear combination of the basis functions $\{\varphi_i\}_{i=1}^N$, so that

$$C_J(q) = \sum_{i=1}^N G_{iJ}\varphi_i(q), \qquad J = 1, \ldots, n_c \tag{7}$$

where each entry G_{iJ} of the matrix $G \in \mathbb{R}^{N \times n_c}$ represents the contribution of the ith basis function to the conformation C_J. By employing a Galerkin discretization of (5), we obtain the stochastic matrix

$$P_{ji}^\tau = \frac{\langle \varphi_j, T^\tau\varphi_i \rangle_\pi}{\langle \varphi_j, \varphi_i \rangle_\pi} = \int_\Omega T^\tau\varphi_i(q) \frac{\varphi_j(q)}{\int_\Omega \varphi_j(q)\pi(q)dq}dq. \tag{8}$$

Taking advantage of the partition of unity property of the set φ_i, we can then localize the global quantities. More precisely, associated with each φ_i, a partial density is given by

$$\pi_i(q) = \frac{\varphi_i(q)\pi(q)}{\int_\Omega \varphi_i(q)\pi(q)}. \tag{9}$$

Note that the term in the denominator of (9) represents the localized thermodynamical weight. Analogous to (6), we then obtain for the coefficients G_{iJ}

$$P^\tau \mathbf{g}_J \approx \mathbf{g}_J, \tag{10}$$

where $\mathbf{g}_J = [G_{1J}, G_{2J}, \ldots, G_{NJ}]^T$.

From the knowledge of the transition matrix P^τ we can identify the metastable states as a linear combination of the meshfree basis functions. This fact is based on the Frobenius Perron theory, which we first introduce for decoupled Markov states in the following section.

2.2   Decoupled Markov States

In the case of a completely decoupled Markov State Model, we do not have metastable states, but n_c stable states. The transition matrix P can be rearranged by permutations such that it has a block diagonal structure, i.e.

$$P = \begin{pmatrix} P_1 & & & \\ & P_2 & & \epsilon \\ & \epsilon & P_3 & \\ & & & \ddots \end{pmatrix} \tag{11}$$

where $\varepsilon = 0$. Each submatrix P_I, $I = 1, \ldots, n_c$ is associated with a right eigenvector u_I and an eigenvalue λ_I, such that $P_I u_I = \lambda_I u_I$. This vector can be trivially embedded such that it is also an eigenvector of the matrix P. For a row stochastic matrix, i.e. $\sum_j P_{ij} = 1$, we can employ the Frobenius – Perron theory, which states that its largest eigenvalue is 1, the corresponding left eigenvector gives the stationary distribution and the elements of the right eigenvector are all identical. Consequently, the matrix P has n_c eigenvalues equal to 1. Thus, the dominant right eigenvectors u_I have the form

$$u_I = (c_1, \ldots, c_1, c_2, \ldots, c_2, c_3, \ldots, c_3, \ldots, c_{n_c}, \ldots, c_{n_c}).$$

Summing up, in a decoupled Markov process, the subspace of the dominant eigenvalue 1 is spanned by n_c piecewise constant eigenvectors. Thus, the metastable states can be identified by the identical entries in the right eigenvectors. Or, alternatively, a linear transformation of the eigenvector basis provides a set of n_c characteristic vectors, such that for the vector u_I corresponding to the I-th cluster all elements are zero except for $c_I = 1$.

2.3 Nearly Decoupled Markov States and Water

In the following, the Markov process is only nearly decoupled, i.e., $\varepsilon > 0$. At the beginning, let us restate the results of the Frobenius-Perron theory from the foregoing section by rewriting G as

$$G = X\mathscr{A}$$

where X represents the eigenvectors of P^τ corresponding to eigenvalues close to one, and the non-singular, still unknown transformation matrix $\mathscr{A} \in \mathbb{R}^{n_c \times n_c}$.

Due to the definition of the conformations and the basis functions, we can state the following constraints:

i. $G_{iJ} \geq 0 \quad \forall i \in \{1, \ldots, N\}, J \in \{1, \ldots, n_c\}$ (positivity)
ii. $\sum_{J=1}^{n_c} G_{iJ} = 1 \quad \forall i \in \{1, \ldots, N\}$ (partition of unity)
iii. $G = X\mathscr{A}$ where $P^\tau X = X\Lambda$, $\Lambda = \mathrm{diag}(\lambda_1, \ldots, \lambda_{n_c})$, $\mathscr{A}$ non-singular (invariance)

The transformation matrix $\mathscr{A}$ is not uniquely defined by these constraints. Given the eigenvectors of P^τ and a transformation matrix $\mathscr{A}$, the matrix elements G_{iJ} are computed according to (iii). The conformations C_I are then constructed according to (7). In order to yield an "optimal" transformation matrix, a corresponding optimization problem is formulated. One possibility is to maximize the "crispness" of the conformations (i.e. $\frac{\langle C_I, C_I \rangle_\pi}{\langle C_I, \mathbf{e} \rangle_\pi}$ close to one), which can be achieved by maximizing

$$I(\mathscr{A}; X, \pi) = \frac{1}{n_c} \sum_{I=1}^{n_c} \frac{\langle C_I, C_I \rangle_\pi}{\langle C_I, \mathbf{e} \rangle_\pi} \leq 1, \tag{12}$$

where $\mathbf{e}$ denotes the constant function equal to 1, cf. [25]. PCCA+ solves the maximization problem (12) subject to the linear constraints (i)–(iii). Given a good starting guess, this convex global maximization problem can be solved by local maximization methods as described in [8, 37].

By PCCA+, the concept of metastable sets is replaced by the concept of metastable membership functions. A conformation of the molecular system is given by a membership function, which may also attain values inside the interval $[0, 1]$. Although this approach seems to be of technical interest only, it is the key to formulate Markov State Models for molecular systems including explicit water molecules. Usually, Markov State Models are said to be invalid in the case of explicit water simulations, because

1. Transitions between molecular conformations can not be modeled as jump-processes, due to the fact that there is a diffusive part in the system. The system is assumed to be non-Markovian.
2. The dynamics of the system is highly non-linear and can not be captured by a linear matrix P.
3. A spectral gap of T^τ for identifying the number of metastable conformations does not exist.

(Ad 1):   A jump inside a Markov State Model, which is based on membership functions, is not a jump between two subsets of the conformational space. Diffusive processes can be modeled correctly by allowing for a "soft" decomposition of the conformational space, cf. [38].

(Ad 2):   Non-linear (stochastic) dynamics corresponds to a linear Fokker-Planck-operator. The main problem is given by the projection error of this linear operator to a low-dimensional transition matrix P. In principle, a correct projection is possible. However, this projection has to include all relevant degrees of freedom which may also include the "conformational state" of the water molecules. Note that only degrees of freedom which are necessary to *distinguish* between different conformations of the system are relevant.

(Ad 3):   In explicit water simulations, the timescales of conformational transitions are not well-separated from the time scales of "internal fluctuations", such that T^τ does not provide a spectral gap. In this situation, the projection error of set-based Markov State Models might be high, because the spectral gap plays an important role in its error estimation, cf. [17]. However, if continuous membership functions are allowed, it is only mandatory to identify an invariant subspace of T^τ, the spectral gap does not enter in the error estimator directly, cf. [27, 38]. Additionally, the spectral gap is not the only possibility to identify the number of clusters in PCCA+, cf. [39].

3 Simulation of Trialanine in an Explicit Solvent

The structure of Trialanine at a pH value of 1 was created by using the visualization software Amira [32]. All molecular dynamics simulations were performed using the molecular dynamics software package GROMACS 4.5 [14]. Trialanine was parameterized according to the Amber-99SB force field [16]. As protonated C-terminal alanine is not part of the standard force field, we created and parameterized a novel residue in compliance with the Amber-99SB force field by using the software Antechamber from AmberTools 1.5 [5,35,36], with charges calculated by the AM1-BCC method. [18, 19]. For the explicit solvent system, we used the TIP4P-Ew water model [15]. Trialanine was placed in a rhombic dodecahedron solvent box of about 4.2 nm side length. In order to neutralize the overall charge, we placed a single negative counter ion in the box. The energy of the system was minimized by using the steepest descent algorithm. Afterwards, a 200 ps simulation was performed during which the position of all heavy atoms of Trialanine was restrained in order to settle the solvent molecules. The output of this run was used as starting configuration for a molecular dynamics run of 100 ns. In order to maintain a constant temperature of 300 K and a pressure of 1 bar, velocity rescaling [4] and Berendsen weak coupling [2] were applied. A twin range cut-off of 1.0/1.4 nm for van-der-Waals interactions was applied and the smooth particle mesh Ewald algorithm [10] was used for Coulomb interactions, with a switching distance of 1.0 nm. Bond length oscillations of bonds involving hydrogen atoms were constrained using the LINCS algorithm [13], allowing for an integration step of 1 fs.

For the ZIBgridfree sampling, the same parameter settings were adopted, save molecular dynamics run time, which was set to 500 ps per node. The ZIBgridfree algorithm uses short local sampling in order to evaluate the partial densities associated with the n basis functions φ_i. In order to ascertain thorough local sampling, each local sampling is confined to the essential support of the corresponding basis function. In this process, the short time trajectories can be parallelized on the force field level (such as provided by state-of-the-art molecular dynamics code). A second level of parallelization is given by the possibility to perform multiple local samplings at the same time, as the basis functions can be evaluated independent of each other. The ZIBgridfree sampling algorithm was implemented using a Python framework built around the GROMACS molecular dynamics code (publication in preparation). In ZIBgridfree, the conformational space discretization is based on internal degrees of freedom. Only degrees of freedom which indicate conformational changes are relevant. Here we follow the work of Mu et al. [22] by taking the two central dihedral angles Φ and Ψ indicated in Fig. 1. In Fig. 2,

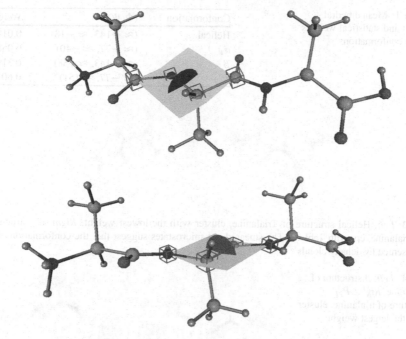

Fig. 1 *Top*: dihedral angle Φ. *Bottom*: dihedral angle Ψ

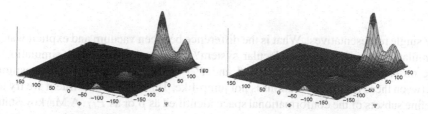

Fig. 2 *Left*: Trialanine in explicit water, computed by molecular dynamics (sampling time 100 ns), *Right*: Trialanine in explicit water, computed by ZIBgridfree (joint sampling time 10 ns)

we plotted the histograms of the two dihedrals in explicit water simulations. The ZIBgridfree simulation shows good agreement with the 100 ns long term molecular dynamics trajectory. It can also be found in literature [9, 21, 22], that the P_{II} (polyglycine II) conformation is predominant in explicit water. The mean dihedral angles and the statistical weights of the four conformations calculated by ZIBgridfree are given in Table 1. However, as observed by Mu et al. [22], the choice of the force field has a dramatic influence on the Boltzmann distribution of the states and thus, a comparison with results from simulations with different force fields is hardly possible. The four identified conformations are given in Figs. 3 and 4. Since we have applied Markov State Models and continuous membership functions in order to identify the conformations, they are given by a family of microstates. Hence, conformations are best represented by density clouds, and not

Table 1 Mean dihedral angles and statistical weights of the conformations

Conformation	(Φ, Ψ)	Weight
Helical	$(\approx -143, \approx -18)$	0.0183
α_R	$(\approx -77, \approx -40)$	0.0583
β	$(\approx -143, \approx 158)$	0.3199
P_{II}	$(\approx -77, \approx 151)$	0.6034

Fig. 3 *Left*: Helical structure of Trialanine, cluster with the lowest weight. *Right*: α_R structure of Trialanine, cp. with Table 1. The smeared-out microstates suggest that the conformations are represented by density clouds

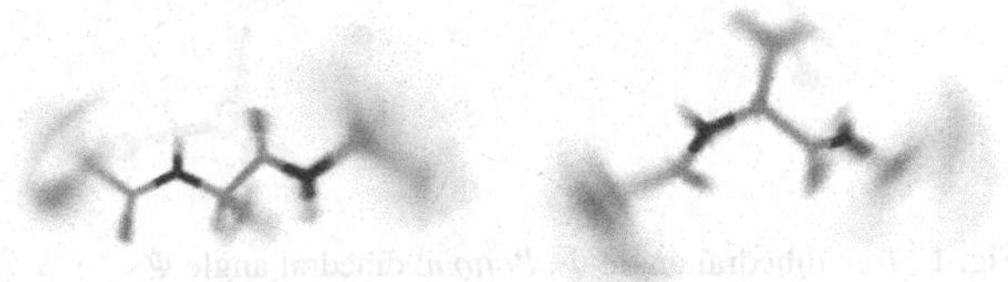

Fig. 4 *Left*: β structure of trialanine. *Right*: P_{II} structure of trialanine, cluster with the largest weight

by single representatives. What is the difference between vacuum and explicit water simulations for the given molecular system? In contrast to vacuum simulations, we observe that the conformations β and P_{II} overlap. Seemingly, the transition between these two conformations is not jump-like. This statement is true if we try to define subsets of the conformational space identified as β or as P_{II}. A Markov State Model based on subsets would lead to a systematic error (and probably to a wrong equilibrium density). However, in the presented approach, subsets are replaced by membership functions. The "Markovian jump" does not take place between two subsets, it takes place between two *overlapping* densities, see also Fig. 5.

For the given example, the spectral gap ($\lambda_4 = 0.97$ and $\lambda_5 = 0.94$) for the calculation of C_I via PCCA+ is not significant as it has been expected for the analysis of an explicit water simulation. Since the membership functions allow for each state a certain "degree of membership", we call these kind of clusters *soft*. The MSM for those membership functions (conformations according to Table 1) is given by

$$P_{\text{soft}}(1ps) = \begin{pmatrix} 0.981 & 0.015 & 0.002 & 0.002 \\ 0.005 & 0.995 & 0.000 & 0.000 \\ 0.000 & 0.000 & 0.982 & 0.018 \\ 0.000 & 0.000 & 0.009 & 0.991 \end{pmatrix},$$

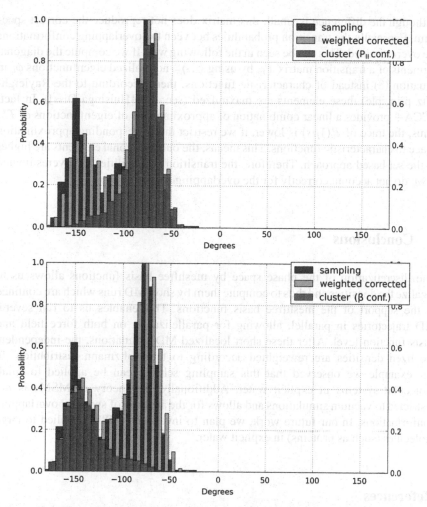

Fig. 5 Distribution of torsion angle Φ. *Dark blue histogram*: Distribution obtained by long term MD (sampling time 100 ns). *Light blue histogram*: Joint reweighted distribution obtained by ZIBgridfree (joint sampling time 10 ns). *Red histogram top*: Density corresponding to Φ-part of the P_{II} conformation. *Red histogram bottom*: Density corresponding to Φ-part of the β conformation

its spectrum exactly reproduces the eigenvalues of the Galerkin discretization of the transfer operator. The construction of a sct-based MSM (*hard cluster*) leads to

$$P_{\text{hard}}(1ps) = \begin{pmatrix} 0.981 & 0.017 & 0.001 & 0.001 \\ 0.007 & 0.993 & 0.000 & 0.000 \\ 0.000 & 0.000 & 0.964 & 0.036 \\ 0.000 & 0.000 & 0.017 & 0.983 \end{pmatrix} .$$

Although the difference is small, this matrix does not reproduce the correct spectrum. Especially the transition probabilities between the overlapping conformations are overestimated. This can be seen in the following way. If we compute the diagonal elements of a transition matrix P_{ij} by using $\langle \cdot, \cdot \rangle_\pi$-normalized eigenfunctions ϕ_i in equation (8) instead of characteristic functions, then (according to the Rayleigh-Ritz principle) these elements are maximized (see also [Huisinga-Diss]). In fact, PCCA+ provides a linear combination of approximations of eigenfunctions of T^τ. Thus, the trace of $P(1ps)$ is lower, if we restrict the corresponding approximation space to characteristic functions. This means, the outer diagonal elements are higher in the set-based approach. Therefore, the transition probabilities are overestimated if we do not account correctly for the overlapping conformations.

4   Conclusions

The discretization of the phase space by meshfree basis functions allows us to localize the densities and thus to compute them by short MD runs which are confined to the support of the meshfree basis functions. This enables us to run several MD trajectories in parallel, allowing for parallelization on both force field and basis function level. After these short localized MD simulations, the independent localized densities are reweighted according to the Boltzmann distribution. In our example we observed that this sampling scheme can be applied to small molecular systems in explicit water. Additionally, the theory of MSMs is not restricted to vacuum simulations and allows for the analysis of spatially overlapping conformations. In our future work, we plan to investigate MSMs applied to large molecules (such as proteins) in explicit water.

References

1. A. Beberg, V.S. Pande, Folding@home: lessons from eight years of distributed computing, in *IEEE Computer Society* Los Alamitos, CA, USA (2009), pp. 1–8
2. H.J.C. Berendsen, J.P.M. Postma, A. DiNola, J.R. Haak, Molecular dynamics with coupling to an external bath. J. Chem. Phys. **81**, 3684–3690 (1984)
3. G.R. Bowman, X. Huang, V.S. Pande Using generalized ensemble simulations and Markov state models to identify conformational states. Methods **49**(2), 197–201 (2009)
4. G. Bussi, D. Donadio, M. Parrinello. Canonical sampling through velocity rescaling. J. Chem. Phys. **126**, 014101 (2007)
5. D.A. Case, T.E. Cheatham III, T. Darden, H. Gohlke, R. Luo, K.M. Merz Jr., A. Onufriev, C. Simmerling, B. Wang, R.J. Woods, The Amber biomolecular simulation programs. J. Comput. Chem. **26**, 1668–1688 (2005)
6. C. Chennubhotla, I. Bahar, Markov methods for hierarchical coarse-graining of large protein dynamics. J. Comput. Biol. **14**(6),765–776, 2007
7. J.D. Chodera, W.C. Swope, J.W. Pitera, K.A. Dill, Long-time protein folding dynamics from short-time molecular dynamics simulations. Multiscale Model. Simul. **5**(4), 1214–1226 (2006)

8. P. Deuflhard, M. Weber, Robust perron cluster analysis in conformation dynamics, in *Linear Algebra and its Applications – Special Issue on Matrices and Mathematical Biology*, ed. by M. Dellnitz, S. Kirkland, M. Neumann, C. Schütte, Vol. 398C (Elsevier, Amsterdam, 2005), pp. 161–184

9. O. Engin, M. Sayar, B. Erman, The introduction of hydrogen bond and hydrophobicity effects into the rotational isomeric states model for conformational analysis of unfolded peptides. Phys. Biol. **6**, 016001 (2009)

10. U. Essmann, L. Perera, M.L. Berkowitz, T. Darden, H. Lee, L.G. Pedersen. A smooth particle mesh Ewald method. J. Chem. Phys. **103**, 8577–8592 (1995)

11. H. Gohlke, M.F. Thorpe, A natural coarse graining for simulating large biomolecular motion. Biophys. J. **91**(6), 2115–2120 (2006)

12. M. Griebel, S. Knapek, G. Zumbusch, *Numerical Simulation in Molecular Dynamics* (Springer, Berlin, Heidelberg, 2007)

13. B. Hess, H. Bekker, H.J.C. Berendsen, J.G.E.M. Fraaije, LINCS: a linear constraint solver for molecular simulations. J. Comput. Chem. **18**, 1463–1472 (1997)

14. B. Hess, C. Kutzner, D. van der Spoel, E. Indahl, GROMACS 4: Algorithms for highly efficient, load-balanced, and scalable molecular simulation. J. Chem. Theory Comput. **4**(3), 435–447 (2008)

15. H.W. Horn, W.C. Swope, J.W. Pitera, Characterization of the TIP4P-Ew water model: vapor pressure and boiling point. J. Chem. Phys. **123**, 194504 (2005)

16. V. Hornak, R. Abel, A. Okur, B. Strockbine, A. Roitberg, C. Simmerling, Comparison of multiple Amber force fields and development of improved protein backbone parameters. Proteins **65**, 712–725 (2006)

17. W. Huisinga, Metastability of Markovian systems: a transfer operator based approach in application to molecular dynamics. Freie Universität, Berlin, Doctoral Thesis, 2001

18. A. Jakalian, B.L. Bush, D.B. Jack, C.I. Bayly, Fast, efficient generation of high-quality atomic charges. AM1-BCC model: I. Method. J. Comput. Chem. **21**, 132–146 (2000)

19. A. Jakalian, D.B. Jack, C.I. Bayly, Fast, efficient generation of high-quality atomic charges. AM1-BCC model: II. Parameterization and validation. J. Comput. Chem. **23**, 1623–1641 (2002)

20. A.R. Leach, *Molecular Modelling: Principles and Applications* (Prentice Hall, Harlow, 2001)

21. Y. Mu, G. Stock, Conformational dynamics of trialanine in water: a molecular dynamics study. J. Phys. Chem. B **106**, 5294–5301 (2002)

22. Y. Mu, D. Kosov, G. Stock, Conformational dynamics of trialanine in Water.2 Comparison of AMBER, CHARMM, GROMOS, and OPLS Force Fields to NMR and infrared experiments J. Phys. Chem. B **107**, 5064– 5073, (2003)

23. A. Pan, B. Roux, Building Markov state models along pathways to determine free energies and rates of transitions. J Chem. Phys. **129**(6), 064107 (2008)

24. J.-H. Prinz, H. Wu, M. Sarich, B. Keller, M. Fischbach, M. Held, J.D. Chodera, C. Schütte, F. Noe, Markov models of molecular kinetics: generation and validation. J. Chem. Phys. **134**, 174105 (2011)

25. S. Röblitz, Statistical error estimation and grid-free hierarchical refinement in conformation dynamics. Doctoral thesis, Department of Mathematics and Computer Science, Freie Universität, Berlin, 2008

26. G. Rose, P. Fleming, J. Banavar, A. Maritan, A backbone-based theory of protein folding. Proc. Natl. Acad. Sci. U. S. A. **103**(45), 16623–16633 (2006)

27. M. Sarich, F. Noe, C. Schütte, On the approximation quality of Markov state models. Multiscale Model. Simul. **8**, 1154–1177 (2010)

28. C. Schütte, Conformational dynamics: modelling, theory, algorithm, and application to biomolecules. Habilitation Thesis, Fachbereich Mathematik und Informatik, Freie Universität Berlin, 1999

29. C. Schütte, A. Fischer, W. Huisinga, P. Deuflhard, A direct approach to conformational dynamics based on hybrid Monte Carlo. J. Comput. Phys. **151**,146–169, 1999. Special Issue Comp. Biophysics, Academic

30. D. Shepard, A two-dimensional interpolation function for irregularly spaced data, in *Proceeding of the 1968 23rd ACM national conference* (ACM, New York, 1968), pp. 517–524
31. S. Sriraman, I. Kevrekidis, G. Hummer, Coarse master equation from Bayesian analysis of replica molecular dynamics simulations. J. Phys. Chem. B **109**(14), 6479–6484 (2005)
32. D. Stalling, M. Westerhoff, H.-C. Hege. Amira: a highly interactive system for visual data analysis, in *The Visualization Handbook*, ed. by C.D. Hansen, C.R. Johnson, Chapter 38 (Elsevier, Amsterdam, 2005) pp. 749–767
33. W.C. Swope, J.W. Pitera, Describing protein folding kinetics by molecular dynamics simulation. 1. Theory. J. Phys. Chem. B **108**, 6571–6581 (2004)
34. W.F. van Gunsteren, H.J.C. Berendsen, Algorithms for macromolecular dynamics and constraint dynamics. Mol. Phys. **34**, 1311–1327 (1977)
35. J. Wang, W. Wang, J. Caldwell, P.A. Kollman, D.A. Case, Development and testing of a general Amber force field. Comput. Chem. **25**, 1157–1174 (2004)
36. J. Wang, W. Wang, P.A. Kollman, D.A. Case, Automatic atom type and bond type perception in molecular mechanical calculations. J. Mol. Graph. Model. **25**, 247–260 (2006)
37. M. Weber, *Meshless Methods in Conformation Dynamics*. Doctoral thesis, Department of Mathematics and Computer Science, Freie Universität, Berlin, 2006. Published by Verlag Dr. Hut, München
38. M. Weber, A subspace approach to molecular Markov state models via a new infinitesimal generator. Habilitation Thesis, Fachbereich Mathematik und Informatik, Freie Universität, Berlin, 2011
39. M. Weber, W. Rungsarityotin, A. Schliep, An indicator for the number of clusters using a linear map to simplex structure, in *From Data and Information Analysis to Knowledge Engineering, Proceedings of the 29th Annual Conference of the Gesellschaft für Klassifikation e.V., Universität Magdeburg, März 2005*, ed. by M. Spiliopoulou et al. Studies in Classification, Data Analysis, and Knowledge Organization (Springer, Berlin, Heidelberg, 2006), pp. 103–110

Kernel-Based Collocation Methods Versus Galerkin Finite Element Methods for Approximating Elliptic Stochastic Partial Differential Equations

Gregory E. Fasshauer and Qi Ye

Abstract We compare a kernel-based collocation method (meshfree approximation method) with a Galerkin finite element method for solving elliptic stochastic partial differential equations driven by Gaussian noises. The kernel-based collocation solution is a linear combination of reproducing kernels obtained from related differential and boundary operators centered at chosen collocation points. Its random coefficients are obtained by solving a system of linear equations with multiple random right-hand sides. The finite element solution is given as a tensor product of triangular finite elements and Lagrange polynomials defined on a finite-dimensional probability space. Its coefficients are obtained by solving several deterministic finite element problems. For the kernel-based collocation method, we directly simulate the (infinite-dimensional) Gaussian noise at the collocation points. For the finite element method, however, we need to truncate the Gaussian noise into finite-dimensional noises. According to our numerical experiments, the finite element method has the same convergence rate as the kernel-based collocation method provided the Gaussian noise is truncated using a suitable number terms.

Keywords Kernel-based collocation • Meshfree approximation • Galerkin finite element • Elliptic stochastic partial differential equations • Gaussian fields • Reproducing kernels

1 Introduction

Stochastic partial differential equations (SPDEs) form the basis of a recent, fast growing research area with many applications in physics, engineering and finance. However, it is often difficult to obtain an explicit form of the solution. Moreover,

G.E. Fasshauer · Q. Ye (✉)
Department of Applied Mathematics, Illinois Institute of Technology, Chicago, IL 60616, USA
e-mail: fasshauer@iit.edu; qye3@iit.edu

M. Griebel and M.A. Schweitzer (eds.), *Meshfree Methods for Partial Differential Equations VI*, Lecture Notes in Computational Science and Engineering 89, DOI 10.1007/978-3-642-32979-1_10, © Springer-Verlag Berlin Heidelberg 2013

current numerical methods usually show limited success for high-dimensional problems and in complex domains. In our recent papers [4, 11], we use a kernel-based collocation method to approximate the solution of high-dimensional SPDE problems. Since parabolic SPDEs can be transformed into elliptic SPDEs using, e.g., an implicit time stepping scheme, solution of the latter represents a particularly important aspect of SPDE problems.

In this paper, we compare the use of a *kernel-based collocation method* [4,11] (meshfree approximation method) and a *Galerkin finite element method* [1, 2] to approximate the solution of elliptic SPDEs. For kernel-based collocation, we directly simulate the Gaussian noise at a set of collocation points. For the Galerkin finite element method, on the other hand, we use a truncated Karhunen-Loéve expansion of the Gaussian noise in order to satisfy a finite-dimensional noise condition. For kernel-based collocation the same collocation locations are used to construct the deterministic basis and the random part. For the Galerkin finite element method one needs to separately set up the finite element basis on the spatial domain and the polynomials on the probability space. For a given kernel function, the convergence rate of the collocation solution depends only on the fill distance of the collocation points. The convergence rate of the finite element solution depends on the maximum mesh spacing parameter and the degrees of the polynomials defined on the finite-dimensional probability space. According to our numerical experiments, the truncation length of the Gaussian noise also affects the convergence results of the finite element method.

1.1 Problem Setting

Assume that $\mathscr{D}$ is a regular open bounded domain in $\mathbb{R}^d$. Let the stochastic process $\xi : \mathscr{D} \times \Omega_\xi \to \mathbb{R}$ be Gaussian with mean zero and covariance kernel $\Phi : \mathscr{D} \times \mathscr{D} \to \mathbb{R}$ defined on a probability space $(\Omega_\xi, \mathscr{F}_\xi, \mathbb{P}_\xi)$ (see Definition 1). We consider an elliptic SPDE driven by the Gaussian noise ξ with Dirichlet boundary conditions

$$\begin{cases} \Delta u = f + \xi, & \text{in } \mathscr{D}, \\ u = 0, & \text{on } \partial\mathscr{D}, \end{cases} \tag{1}$$

where Δ is the Laplacian and $f : \mathscr{D} \to \mathbb{R}$ is a deterministic function. We can solve the SPDE (1) by either of the following two numerical methods.

Kernel-based collocation method (KC): We simulate the Gaussian noise ξ with covariance structure $\Phi(x, y)$ at a finite collection of predetermined *collocation points*

$$X_{\mathscr{D}} := \{x_1, \cdots, x_N\} \subset \mathscr{D}, \quad X_{\partial\mathscr{D}} := \{x_{N+1}, \cdots, x_{N+M}\} \subset \partial\mathscr{D}$$

and approximate u using a kernel-based collocation method written as

$$u(x) \approx \hat{u}(x) := \sum_{k=1}^{N} c_k \Delta_2 \overset{*}{K}(x, x_k) + \sum_{k=1}^{M} c_{N+k} \overset{*}{K}(x, x_{N+k}), \quad x \in \mathscr{D},$$

where $\overset{*}{K}$ is an integral-type kernel associated with a reproducing kernel K (see Eq. (5) in Appendix A.). Here Δ_2 means that we differentiate with respect to the second argument, i.e., $\Delta_2 \overset{*}{K}(x, x_k) = \Delta_y \overset{*}{K}(x, y)|_y = x_k$. The unknown random coefficients $c := (c_1, \cdots, c_{N+M})^T$ are obtained by solving a random system of linear equations (with constant deterministic system matrix and random right-hand side that varies with each realization of the noise). Details are provided in Sect. 2.

Galerkin finite element method (FE): Since the Galerkin finite element method is based on a finite-dimensional noise assumption (see [1, 2]), assuming $\Phi \in L_2(\mathscr{D} \times \mathscr{D})$, we truncate the Gaussian noise ξ by a Karhunen-Loéve expansion, i.e.,

$$\xi \approx \xi^n := \sum_{k=1}^{n} \zeta_k \sqrt{q_k} \phi_k, \quad \text{and} \quad \zeta_k \sim i.i.d. \mathcal{N}(0, 1), \quad k = 1, \cdots, n,$$

where q_k and ϕ_k are eigenvalues and eigenfunctions of the covariance kernel Φ, i.e., $\int_{\mathscr{D}} \Phi(x, y)\phi_k(y)dy = q_k \phi_k(x)$. We approximate the original SPDE (1) by another elliptic SPDE driven by the truncated Gaussian noise ξ^n

$$\begin{cases} \Delta u^n = f + \xi^n, & \text{in } \mathscr{D}, \\ u^n = 0, & \text{on } \partial \mathscr{D}. \end{cases} \tag{2}$$

Next we combine the finite element method in spatial domain $\overline{\mathscr{D}} := \mathscr{D} \cup \partial \mathscr{D}$ and the collocation in the zeros of suitable tensor product orthogonal polynomials (Gaussian points) in the finite-dimensional probability space. We obtain the approximation as a tensor product of the finite element solutions defined on the spatial domain and the Lagrange polynomials defined on the finite-dimensional probability space, i.e., $u_{h,p} \approx u^n \approx u$, where h is the maximum mesh spacing parameter and $p = (p_1, \cdots, p_n)$ is the degree of the Lagrange polynomials. Details are provided in Sect. 3.

Remark 1. Because Φ is always positive semi-definite Mercer's theorem ensures that its eigenvalues $q_1 \geq q_2 \geq \cdots \geq 0$ and its eigenfunctions $\{\phi_k\}_{k=1}^{\infty}$ form an orthonormal base of $L_2(\mathscr{D})$ so that $\Phi(x, y) = \sum_{k=1}^{\infty} q_k \phi_k(x)\phi_k(y)$. Therefore $E \|\xi - \xi^n\|_{L_2(\mathscr{D})}^2 = \sum_{k=n+1}^{\infty} q_k \to 0$ when $n \to \infty$, and we can accurately represent the infinite-dimensional noise ξ by a (potentially long) truncated Karhunen-Loéve expansion.

2 Kernel-Based Collocation Method

In the papers [4, 11] we use the Gaussian fields ΔS, S with means $\Delta \mu$, μ and covariance kernels $\Delta_1 \Delta_2 \overset{*}{K}$, $\overset{*}{K}$ (see Theorem 1), respectively, to construct the collocation approximation $\hat{u}$ of the solution u of SPDE (1). Here $\Delta_1 \Delta_2 \overset{*}{K}(\boldsymbol{x}_j, \boldsymbol{x}_k) = \Delta_{\boldsymbol{x}} \Delta_{\boldsymbol{y}} \overset{*}{K}(\boldsymbol{x}, \boldsymbol{y})|_{\boldsymbol{x}=\boldsymbol{x}_j, \boldsymbol{y}=\boldsymbol{x}_k}$.

Because of the order $\mathcal{O}(\Delta) = 2$, we suppose that the reproducing kernel Hilbert space $\mathrm{H}_K(\mathscr{D})$ is embedded into the L_2-based Sobolev space $\mathscr{H}^m(\mathscr{D})$ where $m > 2 + d/2$.

Remark 2. Since we want to interpolate the values of the differential equation at the collocation points, $\Delta \omega(\boldsymbol{x})$ needs to be well-defined pointwise for each available solution $\mathscr{D} \in \mathrm{H}_K(\mathscr{D}) \subseteq \mathscr{H}^m(\Omega) \subset \mathrm{C}^2(\overline{\mathscr{D}})$. This requires the Sobolev space $\mathscr{H}^m(\mathscr{D})$ to be smooth enough. If we just need a weak solution as for the finite element method, then the order needs to satisfy $m \geq 2$ only.

Since ξ is Gaussian with a known correlation structure, we can simulate the values of ξ at the collocation points $\boldsymbol{x}_1, \cdots, \boldsymbol{x}_N$, i.e.,

$$\boldsymbol{\xi} := (\xi_{\boldsymbol{x}_1}, \cdots, \xi_{\boldsymbol{x}_N})^T \sim \mathcal{N}(\boldsymbol{0}, \boldsymbol{\Phi}), \quad \text{where } \boldsymbol{\Phi} := (\Phi(\boldsymbol{x}_j, \boldsymbol{x}_k))_{j,k=1}^{N,N}.$$

Consequently, we assume that the values $\{y_j\}_{j=1}^N$ and $\{y_{N+j}\}_{j=1}^M$ defined by

$$y_j := f(\boldsymbol{x}_j) + \xi_{\boldsymbol{x}_j}, \quad j = 1, \cdots, N, \quad y_{N+j} := 0, \quad j = 1, \cdots, M,$$

are known. Moreover we can also obtain the joint probability density function p_y of the random vector $\boldsymbol{y}_\xi := (y_1, \cdots, y_{N+M})^T$.

We define the product space

$$\Omega_{K\xi} := \Omega_K \times \Omega_\xi, \quad \mathscr{F}_{K\xi} := \mathscr{F}_K \otimes \mathscr{F}_\xi, \quad \mathbb{P}_\xi^\mu := \mathbb{P}^\mu \otimes \mathbb{P}_\xi,$$

where the probability measure $\mathbb{P}^\mu$ is defined on $(\mathrm{H}_K(\mathscr{D}), \mathscr{B}(\mathrm{H}_K(\mathscr{D}))) = (\Omega_K, \mathscr{F}_K)$ as in Theorem 1, and the probability space $(\Omega_\xi, \mathscr{F}_\xi, \mathbb{P}_\xi)$ is given in the SPDE (1). We assume that the random variables defined on the original probability spaces are extended to random variables on the new probability space in the natural way: if random variables $V_1 : \Omega_K \to \mathbb{R}$ and $V_2 : \Omega_\xi \to \mathbb{R}$ are defined on $(\Omega_K, \mathscr{F}_K, \mathbb{P}^\mu)$ and $(\Omega_\xi, \mathscr{F}_\xi, \mathbb{P}_\xi)$, respectively, then

$$V_1(\omega_1, \omega_2) := V_1(\omega_1), \ V_2(\omega_1, \omega_2) := V_2(\omega_2), \quad \text{for each } \omega_1 \in \Omega_K \text{ and } \omega_2 \in \Omega_\xi.$$

Note that in this case the random variables have the same probability distributional properties, and they are independent on $(\Omega_{K\xi}, \mathscr{F}_{K\xi}, \mathbb{P}_\xi^\mu)$. This implies that the stochastic processes ΔS, S and ξ can be extended to the product space

$(\Omega_{K\xi}, \mathscr{F}_{K\xi}, \mathbb{P}_{\xi}^{\mu})$ while preserving the original probability distributional properties, and that $(\Delta S, S)$ and ξ are independent.

2.1 Approximation of SPDEs

Fix any $x \in \mathscr{D}$. Let $\mathscr{A}_x(v) := \{\omega_1 \times \omega_2 \in \Omega_{K\xi} : \omega_1(x) = v\}$ for each $v \in \mathbb{R}$, and $\mathscr{A}_{PB}^{y_\xi} := \{\omega_1 \times \omega_2 \in \Omega_{K\xi} : \Delta\omega_1(x_1) = y_1(\omega_2), \ldots, \omega_1(x_{N+M}) = y_{N+M}(\omega_2)\}$. Using the methods in [4] and Theorem 1, we obtain

$$\mathbb{P}_{\xi}^{\mu}(\mathscr{A}_x(v)|\mathscr{A}_{PB}^{y_\xi}) = \mathbb{P}_{\xi}^{\mu}(S_x = v|S_{PB} = y_{\xi}) = p_x^{\mu}(v|y_{\xi}),$$

where $p_x^{\mu}(\cdot|\cdot)$ is the conditional probability density function of the random variable S_x given the random vector $S_{PB} := (\Delta S_{x_1}, \cdots, \Delta S_{x_N}, S_{x_{N+1}}, \cdots, S_{x_{N+M}})^T$. (Here y_{ξ} is viewed as given values.) According to the natural extension rule, p_x^{μ} is consistent with the formula (6). Then the approximation $\hat{u}(x)$ is solved by the maximization problem

$$\hat{u}(x) = \underset{v\in\mathbb{R}}{\arg\max} \ \underset{\mu\in H_K(\mathscr{D})}{\sup} \ p_x^{\mu}(v|y_{\xi}).$$

If the covariance matrix

$$\overset{*}{K}_{PB} := \begin{pmatrix} (\Delta_1\Delta_2\overset{*}{K}(x_j,x_k))_{j,k=1}^{N,N}, & (\Delta_1\overset{*}{K}(x_j,x_{N+k}))_{j,k=1}^{N,M} \\ (\Delta_2\overset{*}{K}(x_{N+j},x_k))_{j,k=1}^{M,N}, & (\overset{*}{K}(x_{N+j},x_{N+k}))_{j,k=1}^{M,M} \end{pmatrix} \in \mathbb{R}^{(N+M)\times(N+M)}$$

is nonsingular, then one solution of the above maximum problem has the form

$$\hat{u}(x) := \sum_{k=1}^{N} c_k \Delta_2\overset{*}{K}(x,x_k) + \sum_{k=1}^{M} c_{N+k}\overset{*}{K}(x,x_{N+k}) = k_{PB}(x)^T \overset{*}{K}_{PB}^{-1} y_{\xi}, \quad (3)$$

where $k_{PB}(x) := (\Delta_2\overset{*}{K}(x,x_1), \cdots, \Delta_2\overset{*}{K}(x,x_N), \overset{*}{K}(x,x_{N+1}), \cdots, \overset{*}{K}(x,x_{N+M}))^T$. This means that its random coefficients are obtained from the linear equation system $\overset{*}{K}_{PB}c = y_{\xi}$.

The estimator $\hat{u}$ also satisfies the interpolation condition, i.e., $\Delta\hat{u}(x_1) = y_1, \ldots, \Delta\hat{u}(x_N) = y_N$ and $\hat{u}(x_{N+1}) = y_{N+1}, \ldots, \hat{u}(x_{N+M}) = y_{N+M}$. It is obvious that $\hat{u}(\cdot, \omega_2) \in H_K(\mathscr{D})$ for each $\omega_2 \in \Omega_{\xi}$. Since the random part of $\hat{u}(x)$ is only related to y_{ξ}, we can formally rewrite $\hat{u}(x, \omega_2)$ as $\hat{u}(x, y_{\xi})$ and $\hat{u}(x)$ can be transferred to a random variable defined on the finite-dimensional probability space $(\mathbb{R}^{N+M}, \mathscr{B}(\mathbb{R}^{N+M}), \mu_y)$, where the probability measure μ_y is defined by $\mu_y(dv) := p_y(v)dv$. Moreover, the probability distributional properties of $\hat{u}(x)$ do not change when $(\Omega_{\xi}, \mathscr{F}_{\xi}, \mathbb{P}_{\xi})$ is replaced by $(\mathbb{R}^{N+M}, \mathscr{B}(\mathbb{R}^{N+M}), \mu_y)$.

Remark 3. The random coefficients are obtained solving by system of linear equations that is slightly different from [4]. However the main ideas and techniques are the same as in [4]. For this estimator it is easier to derive error bounds and compare with Galerkin finite element method. A lot more details of the relationship between the two different estimators are provided in [11, Chap. 7].

2.2   Convergence Analysis

We assume that $u(\cdot, \omega_2)$ belongs to $H_K(\mathscr{D})$ almost surely for $\omega_2 \in \Omega_\xi$. Therefore u can be seen as a map from Ω_ξ into $H_K(\mathscr{D})$. So we have $u \in \Omega_{K\xi} = \Omega_K \times \Omega_\xi$.

We fix any $x \in \mathscr{D}$ and any $\epsilon > 0$. Let the subset

$$\mathscr{E}_x^\epsilon := \Big\{ \omega_1 \times \omega_2 \in \Omega_{K\xi} : |\omega_1(x) - \hat{u}(x, \omega_2)| \geq \epsilon,$$

$$\text{such that } \Delta\omega_1(x_1) = y_1(\omega_2), \dots, \omega_1(x_{N+M}) = y_{N+M}(\omega_2) \Big\}.$$

Because $\Delta S_x(\omega_1, \omega_2) = \Delta S_x(\omega_1) = \Delta\omega_1(x)$, $S_x(\omega_1, \omega_2) = S_x(\omega_1) = \omega_1(x)$ and $y_\xi(\omega_1, \omega_2) = y_\xi(\omega_2)$ for each $\omega_1 \in \Omega_K$ and $\omega_2 \in \Omega_\xi$ (see Theorem 1) we can deduce that

$$\mathbb{P}_\xi^\mu \left(\mathscr{E}_x^\epsilon \right) = \mathbb{P}_\xi^\mu \left(|S_x - \hat{u}(x)| \geq \epsilon \text{ such that } S_{PB} = y_\xi \right)$$

$$= \int_{\mathbb{R}^{N+M}} \int_{|v-\hat{u}(x,v)| \geq \epsilon} p_x^\mu(v|v) p_y(v) \mathrm{d}v \mathrm{d}v$$

$$= \int_{\mathbb{R}^{N+M}} \mathrm{erfc}\left(\frac{\epsilon}{\sqrt{2}\sigma(x)} \right) p_y(v) \mathrm{d}v = \mathrm{erfc}\left(\frac{\epsilon}{\sqrt{2}\sigma(x)} \right),$$

where the variance of p_x^μ is $\sigma(x)^2 = \overset{*}{K}(x,x) - k_{PB}(x)^T \overset{*}{K}_{PB}^{-1} k_{PB}(x)$ (see Eq. (6) given in Appendix A.).

The reader may note that the form of the expression for the variance $\sigma(x)^2$ is analogous to that of the *power function* [5, 10], and we can therefore use the same techniques as in the proofs from [4, 5, 10, 11] to obtain a formula for the order of $\sigma(x)$, i.e.,

$$\sigma(x) = \mathscr{O}(h_X^{m-2-d/2}),$$

where $h_X = \sup_{x \in \mathscr{D}} \min_{x_j \in X_\mathscr{D} \cup X_{\partial\mathscr{D}}} \| x - x_j \|_2$ is the *fill distance* of $X := X_\mathscr{D} \cup X_{\partial\mathscr{D}}$. This implies that

$$\sup_{\mu \in H_K(\mathscr{D})} \mathbb{P}_\xi^\mu(\mathscr{E}_x^\epsilon) = \mathscr{O}\left(\frac{h_X^{m-2-d/2}}{\epsilon} \right).$$

Because $|u(x, \omega_2) - \hat{u}(x, \omega_2)| \geq \epsilon$ if and only if $u \in \mathscr{E}_x^\epsilon$ we conclude that

$$\sup_{\mu \in H_K(\mathscr{D})} \mathbb{P}_\xi^\mu \left(\|u - \hat{u}\|_{L_\infty(\mathscr{D})} \geq \epsilon \right) \leq \sup_{\mu \in H_K(\mathscr{D}), x \in \mathscr{D}} \mathbb{P}_\xi^\mu \left(\mathscr{E}_x^\epsilon \right) \to 0, \quad \text{when } h_X \to 0.$$

Therefore we say that the estimator $\hat{u}$ converges to the exact solution u of the SPDE (1) in all probabilities $\mathbb{P}_\xi^\mu$ when h_X goes to 0.

Sometimes we know only that the solution $u \in \mathscr{H}^m(\mathscr{D})$. In this case, as long as the reproducing kernel Hilbert space is dense in the Sobolev space $\mathscr{H}^m(\mathscr{D})$ with respect to its Sobolev norm, we can still say that $\hat{u}$ converges to u in probability.

3 Galerkin Finite Element Method

The right hand side of the SPDE (2)

$$f_{\xi^n}(x, \zeta) := f(x) + \xi_x^n = f(x) + \sum_{k=1}^{n} \zeta_k \sqrt{q_k} \phi_k(x), \quad x \in \mathscr{D},$$

and the random vector $\zeta := (\zeta_1, \cdots, \zeta_n)^T$ has the joint standard normal density function

$$\rho_n(z) := \prod_{k=1}^{n} \rho(z_k), \quad z \in \mathbb{R}^n, \quad \text{where } \rho(z) := \frac{1}{\sqrt{2\pi}} e^{-z^2/2}.$$

Therefore we can replace the probability space $(\Omega_\xi, \mathscr{F}_\xi, \mathbb{P}_\xi)$ by a finite-dimensional probability space $(\mathbb{R}^n, \mathscr{B}(\mathbb{R}^n), \mu_\zeta)$ such that u^n and ξ^n have the same probability distributional properties on both probability spaces, where the probability measure μ_ζ is defined by $\mu_\zeta(\mathrm{d}z) := \rho_n(z)\mathrm{d}z$.

In the paper [1] the numerical approximation $u_{h,p}$ of the solution u^n of the SPDE (2) is sought in a finite-dimensional subspace $V_{h,p}$ based on a tensor product, $V_{h,p} := H_h(\mathscr{D}) \otimes \mathscr{P}_p(\mathbb{R}^n)$, where the following hold:

(C1) $H_h(\mathscr{D}) \subset \mathscr{H}_0^1(\mathscr{D})$ is a standard finite element space, which contains continuous piecewise polynomials defined on regular triangulations with a maximum mesh spacing parameter h.

(C2) $\mathscr{P}_p(\mathbb{R}^n) := \otimes_{k=1}^n \mathscr{P}_{p_k}(\mathbb{R}) \subset L_{2,\rho_n}(\mathbb{R}^n)$ is the span of the tensor product of polynomials with degree at most $p = (p_1, \cdots, p_n)$, where $\mathscr{P}_{p_k}(\mathbb{R})$ is a space of univariate polynomials of degree p_k for each $k = 1, \ldots, n$.

Thus the approximation $u_{h,p} \in V_{h,p}$ and $u_{h,p}(x)$ is a random variable defined on the finite-dimensional probability space $(\mathbb{R}^n, \mathscr{B}(\mathbb{R}^n), \mu_\zeta)$.

Next we construct the Gaussian points

$$z_j := (z_{1,j_1}, \cdots, z_{n,j_n})^T, \quad j \in \mathcal{N}_p := \{j \in \mathbb{N}^n : 1 \leq j_k \leq p_k + 1, \, k = 1, \ldots, n\},$$

where $z_{k,1}, \ldots, z_{k,p_k+1}$ are the roots of the Hermite polynomials η_{p_k+1} of degree $p_k + 1$ for each dimension $k = 1, \ldots, n$. The η_{p_k+1} are also orthogonal polynomials on the space $\mathcal{P}_{p_k}(\mathbb{R})$ with respect to a standard normal weight ρ. Here these Hermite polynomials are only used to set up the Gaussian points for approximating the Gaussian fields.

Let a polynomial base of $\mathcal{P}_p(\mathbb{R}^n)$ be

$$l_j(z) := \prod_{k=1}^n l_{k,j_k}(z_k), \quad j \in \mathcal{N}_p,$$

where $\{l_{k,j}\}_{j=1}^{p_k+1}$ is the Lagrange basis of $\mathcal{P}_{p_k}(\mathbb{R})$ for each $k = 1, \ldots, n$, i.e.,

$$l_{k,j} \in \mathcal{P}_{p_k}(\mathbb{R}), \quad l_{k,j}(z_{k,i}) = \delta_{ij}, \quad i, j = 1, \ldots, p_k + 1,$$

and δ_{ij} is the Kronecker symbol.

For each Gaussian point $z_j \in \mathcal{N}_p$, we compute the finite element solution $u_h(\cdot, z_j) \in H_h(\mathscr{D})$ of the equation

$$-\int_{\mathscr{D}} \nabla u_h(x, z_j) \nabla \gamma(x) dx = \int_{\mathscr{D}} f_{\xi^n}(x, z_j) \gamma(x) dx, \quad \text{for any } \gamma \in H_h(\mathscr{D}).$$

The approximation $u_{h,p}$ is the tensor product of the finite element solutions and the Lagrange polynomials, i.e.,

$$u_{h,p}(x, \zeta) := \sum_{j \in \mathcal{N}_p} u_h(x, z_j) l_j(\zeta), \quad x \in \mathscr{D}. \tag{4}$$

This indicates that $u_{h,p}$ is interpolating at all Gaussian points $z_j \in \mathcal{N}_p$.

We assume that u^n belongs to $L_{2,\rho_n}(\mathbb{R}^n) \otimes \mathscr{H}_0^1(\mathscr{D})$. According to [1, Theorem 4.1] and [1, Lemma 4.7] we get the error bound

$$\left\| E_{\rho_n}(u^n - u_{h,p}) \right\|_{\mathscr{H}_0^1(\mathscr{D})}$$

$$\leq C_1 \inf_{w \in L_{2,\rho_n}(\mathbb{R}^n) \otimes H_h(\mathscr{D})} \left(\int_{\mathscr{D}} E_{\rho_n} |\nabla u^n(x) - \nabla w(x)|^2 dx \right)^{1/2} + C_2 \sum_{k=1}^n p_k^{3/2} e^{-r_k p_k^{1/2}}$$

with positive constants $r_1, \ldots, r_n$ and C_1, C_2 independent of h and p.

4 Side-by-Side Comparison of Both Methods

4.1 Differences

- **Probability spaces:** For kernel-based collocation (KC) we transfer the probability space $(\Omega_\xi, \mathscr{F}_\xi, \mathbb{P}_\xi)$ to the tensor product probability space $(\Omega_{K\xi}, \mathscr{F}_{K\xi}, \mathbb{P}_\xi^\mu)$ such that the Gaussian noise ξ has the same probability distributional properties defined on both probability spaces, while for Galerkin finite elements (FE) we approximate the Gaussian noise ξ by the truncated Gaussian noise ξ^n such that $\lim_{n\to\infty} E \|\xi - \xi^n\|_{\mathrm{L}_2(\mathscr{D})} = 0$ and ξ^n has the same probability distributional properties on the probability space $(\Omega_\xi, \mathscr{F}_\xi, \mathbb{P}_\xi)$ and the finite-dimensional probability space $(\mathbb{R}^n, \mathscr{B}(\mathbb{R}^n), \mu_\xi)$

- **Basis functions:** The bases of the KC solution $\hat{u}$ are the kernel functions $\Delta_2 \overset{*}{K}$ and $\overset{*}{K}$ centered at the collocation points $X_{\mathscr{D}} \subset \mathscr{D}$ and $X_{\partial\mathscr{D}} \subset \partial\mathscr{D}$, while the bases of the FE solution $u_{h,p}$ are the tensor products of the triangular finite element bases defined on $\mathscr{D}$ and the Lagrange polynomials defined on $(\mathbb{R}^n, \mathscr{B}(\mathbb{R}^n), \mu_\xi)$.

- **Simulation:** For KC we can simulate the Gaussian noise ξ at the collocation points $X_{\mathscr{D}}$ because we know its covariance kernel Φ, i.e., $\xi = (\xi_{x_1}, \cdots, \xi_{x_N})^T \sim \mathscr{N}(\mathbf{0}, \Phi)$ and $\Phi = (\Phi(x_j, x_k))_{j,k=1}^{N,N}$. For FE we can simulate the random vector $\zeta = (\zeta_1, \cdots, \zeta_2)^T \sim \mathscr{N}(\mathbf{0}, I_n)$ in order to introduce random variables on $(\mathbb{R}^n, \mathscr{B}(\mathbb{R}^n), \mu_\zeta)$.

- **Interpolation:** In KC $\Delta\hat{u}$ and $\hat{u}$ are interpolating at the collocation points $X_{\mathscr{D}} \cup X_{\partial\mathscr{D}} \subset \mathscr{D}$ in the domain space, respectively, while in FE $u_{h,p}$ is interpolating at the Gaussian points $\mathscr{N}_p \subset \mathbb{R}^n$ in the probability space.

- **Function spaces:** For KC, $\hat{u} \in \mathrm{span}\{\Delta_2 \overset{*}{K}(\cdot, x_j), \overset{*}{K}(\cdot, x_{N+k})\}_{j,k=1}^{N,M} \otimes \mathscr{P}_1$ $(\mathbb{R}^{N+M}) \subset \mathrm{H}_K(\mathscr{D}) \otimes \mathrm{L}_{2,p_y}(\mathbb{R}^{N+M})$, while for FE we have $u_{h,p} \in \mathrm{H}_h(\mathscr{D}) \otimes \mathscr{P}_p(\mathbb{R}^n) \subset \mathscr{H}_0^1(\mathscr{D}) \otimes \mathrm{L}_{2,\rho_n}(\mathbb{R}^n)$.

- **Approximation properties:** The KC result $\hat{u}$ approximates the solution u of the SPDE (1) and its convergence rate depends on the fill distance h_X of the collocation points, while the FE result $u_{h,p}$ approximates the truncated solution u^n of the SPDE (2) with a convergence rate that depends on the maximum mesh spacing parameter h of the triangulation and the degree p of the Lagrange polynomials.

4.2 Relationship Between the Two Methods

Roughly speaking, the random parts of $\hat{u}$ and $u_{h,p}$ are simulated by the normal random vectors ξ and ζ, respectively.

For the following we assume that Φ is positive definite on $\mathcal{D}$ and the dimensions of ξ and ζ are the same, i.e., $N = n$.

We firstly show the relationship between ξ and ζ. Since Φ is positive definite, we have the decomposition $\Phi = \mathsf{V}\mathsf{D}\mathsf{V}^T$, where D and V are the eigenvalues and eigenvector matrices of Φ, respectively. Therefore

$$\xi \sim \mathsf{V}\mathsf{D}^{1/2}\zeta \sim \mathcal{N}(\mathbf{0}, \Phi), \quad \zeta \sim \mathsf{D}^{-1/2}\mathsf{V}^T\xi \sim \mathcal{N}(\mathbf{0}, I_N).$$

We can also use ξ and ζ to approximate the Gaussian noise ξ_x for any fixed $x \in \mathcal{D}$. Using simple kriging, we let

$$\hat{\xi}_x := c(x)^T\xi = b(x)^T\Phi^{-1}\xi \sim \mathcal{N}(0, b(x)^T\Phi^{-1}b(x)),$$

where $b(x) := (\Phi(x, x_1), \cdots, \Phi(x, x_N))^T$. According to [9],

$$E\left|\xi_x - \hat{\xi}_x\right|^2 = \Phi(x, x) - b(x)^T\Phi^{-1}b(x) = \mathcal{O}(h_{X_{\mathcal{D}}}^k),$$

when $\Phi \in C^{2k}(\mathcal{D} \times \mathcal{D})$ and $h_{X_{\mathcal{D}}}$ is the fill distance of $X_{\mathcal{D}}$. For the Karhunen-Loéve expansion,

$$\xi_x^N = \sum_{j=1}^{N} \sqrt{q_j}\phi_j(x)\zeta_j = \phi(x)^T\mathsf{Q}^{1/2}\zeta \sim \mathcal{N}(0, \phi(x)^T\mathsf{Q}\phi(x)),$$

where $\phi(x) := (\phi_1(x), \cdots, \phi_N(x))^T$ and $\mathsf{Q} = \mathrm{diag}(q_1, \cdots, q_N)$. We also have

$$E\int_{\mathcal{D}}\left|\xi_x - \xi_x^N\right|^2 \mathrm{d}x = \sum_{j=N+1}^{\infty} q_j.$$

This shows that the kernel-based method and the finite element method, respectively, use a kernel basis and a spectral basis to approximate Gaussian fields. It also shows that we should suitably choose collocation points such that

$$\lim_{N\to\infty} h_{X_{\mathcal{D}}} = 0 \quad \Longrightarrow \quad \lim_{N\to\infty}\hat{\xi} = \lim_{N\to\infty}\xi^N = \xi.$$

Usually, the smoothness of Φ is related to its eigenvalues in the sense that the order k of continuous differentiability becomes large when the eigenvalues q_j decrease fast, e.g., $\Phi(x, y) = \sum_{j=1}^{\infty}(2\pi j)^{-2k}\sin(2\pi jx)\sin(2\pi jy)$.

Following these discussions, when the eigenvalues of the covariance kernel Φ decrease fast, then the Galerkin finite element method seems to be preferable to the kernel-based collocation method because we can truncate the Gaussian noise ξ at a low dimension. However, when the eigenvalues change slowly, then we may use the kernel-based collocation method because we are able to directly simulate ξ by its covariance structure.

4.3   Competitive Advantages

- The kernel-based collocation method is a meshfree approximation method. It does not require an underlying triangular mesh as the Galerkin finite element method does. For both methods, the (collocation) points can be placed at rather arbitrarily scattered locations which allows for the use of either deterministic or random designs, e.g., Halton or Sobol' points.
- The kernel-based collocation method can be applied to a high-dimensional domain $\mathscr{D}$ with complex boundary $\partial\mathscr{D}$. We can also generalize it to solve a system of elliptic SPDEs derived by vector Gaussian noises.
- The collocation method requires the SPDE solution to be smooth enough such that interpolation is well-behaved at each collocation point while the finite element method can solve non-smooth problems.
- The interpolation matrix $\overset{*}{K}_{PB}$ for the collocation method is usually a dense (and sometimes ill-conditioned) matrix. The finite element method, on the other hand, usually gets the solutions by a sparse linear system because its basis consists of local elements.
- To obtain the truncated Gaussian noise ξ^n for the finite element method we need to compute the eigenvalues and eigenfunctions of the covariance kernel Φ. This, however, is usually difficult to do, so one must estimate them. For the collocation method we need not worry about this issue.
- If the truncated dimension n for the finite element solutions is large, then the degree p of the polynomials has to become correspondingly large in order to satisfy a given error tolerance (see [2]). Once the kernel functions are fixed, the error of the collocation solution only depends on the collocation points.
- In the finite element method, the dimension of its polynomial space defined on the finite-dimensional probability space is equal to $n_p = \prod_{k=1}^{n}(p_k + 1)$. So we need to compute n_p deterministic finite element solutions. In the collocation method, we need to simulate the N-dimensional nonstandard normal vector. Therefore, when $n \ll N$ we may choose the finite element method, while vice versa the collocation method may be preferable.
- For various covariance kernels Φ for the Gaussian noise ξ, the choice of reproducing kernels can affect the kernel-based collocation solution. How to choose the "best" kernels is still an open question. Polynomials are used to construct the approximations for the finite element method, and we only need to determine the appropriate polynomial degree.
- The paper [1] also discusses any other tensor-product finite dimensional noise. In the papers [4, 11] we only consider Gaussian noises. However, one may generalize the kernel-based collocation idea to other problems with colored noise.
- The finite element method works for any elliptic SPDE whose differential operator contains stochastic coefficients defined on a finite dimensional probability space. For the collocation method this idea requires further study.

5   Numerical Examples

In this section we present a few simple numerical experiments comparing the kernel-based collocation method to the Galerkin finite element method.

Let the domain $\mathscr{D} := (0,1)^2 \subset \mathbb{R}^2$ and the covariance kernel of the (finite-dimensional) noise be

$$\Phi(x,y) := 4\pi^4 \sin(\pi x_1) \sin(\pi x_2) \sin(\pi y_1) \sin(\pi y_2)$$
$$+ 16\pi^4 \sin(2\pi x_1) \sin(2\pi x_2) \sin(2\pi y_1) \sin(2\pi y_2)$$

so that we are able to demonstrate the effects of a "correct" and "incorrect" truncation dimension for the finite element method. We use the deterministic function

$$f(x) := -2\pi^2 \sin(\pi x_1) \sin(\pi x_2) - 8\pi^2 \sin(2\pi x_1) \sin(2\pi x_2)$$

and the Gaussian noise ξ with the covariance kernel Φ to set up the right hand side of the stochastic Poisson equation with Dirichlet boundary condition as in SPDE (1). Then its solution has the form

$$u(x) := \sin(\pi x_1) \sin(\pi x_2) + \sin(2\pi x_1) \sin(2\pi x_2) + \zeta_1 \sin(\pi x_1) \sin(\pi x_2)$$

$$+ \frac{\zeta_2}{2} \sin(2\pi x_1) \sin(2\pi x_2), \quad x = (x_1, x_2) \in \mathscr{D},$$

where ζ_1, ζ_2 are independent standard normal random variables defined on $(\Omega_\xi, \mathscr{F}_\xi, \mathbb{P}_\xi)$, i.e., $\zeta_1, \zeta_2 \sim$ i.i.d. $\mathcal{N}(0,1)$.

For the collocation methods, we use the C^4-Matérn function with shape parameter $\theta > 0$

$$g_\theta(r) := (3 + 3\theta r + \theta^2 r^2)e^{-\theta r}, \quad r > 0,$$

to construct the reproducing kernel (Sobolev-spline kernel)

$$K_\theta(x,y) := g_\theta(\|x - y\|_2), \quad x, y \in \mathscr{D}.$$

According to [6], we can deduce that its reproducing kernel Hilbert space $H_K(\mathscr{D})$ is equivalent to the L_2-based Sobolev space $\mathscr{H}^{3+1/2}(\mathscr{D}) \subset C^2(\overline{\mathscr{D}})$. Then we can compute the integral-type $\overset{*}{K}_\theta(x,y) = \int_0^1 \int_0^1 K_\theta(x,z)K_\theta(y,z)dz_1dz_2$. Next we choose Halton points in $\mathscr{D}$ and evenly spaced points on $\partial\mathscr{D}$ as collocation points. Using the kernel-based collocation method, we can set up the approximation $\hat{u}$ via formula (3).

Since we chose $\xi_x = \zeta_1 2\pi^2 \sin(\pi x_1) \sin(\pi x_2) + \zeta_2 4\pi^2 \sin(2\pi x_1) \sin(2\pi x_2)$, we can let the dimension of the probability space be either $n = 1$ or $n = 2$ for the finite element method. The Gaussian points $z_{j,1}, \cdots, z_{j,p_j+1} \in \mathbb{R}$ are computed as the roots of the Hermite polynomial of degree $p_j + 1$, $j = 1, 2$. For $n = 1$ we have $\xi^1 = \zeta_1 2\pi^2 \sin(\pi x_1) \sin(\pi x_2)$, while $n = 2$ gives $\xi^2 = \xi$.

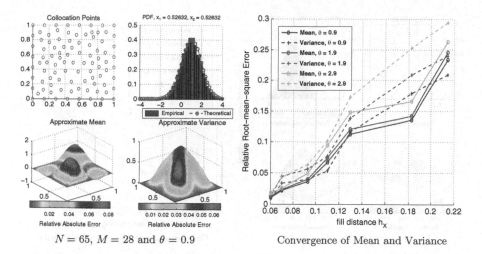

$$N = 65, \ M = 28 \text{ and } \theta = 0.9$$

Convergence of Mean and Variance

Fig. 1 Kernel-based collocation method

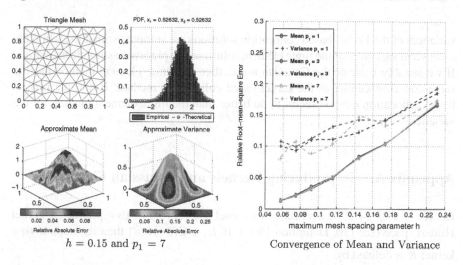

$$h = 0.15 \text{ and } p_1 = 7$$

Convergence of Mean and Variance

Fig. 2 Galerkin finite element methods, $n = 1$

We approximate the mean and variance of the arbitrary random variable U by its sample mean and sample variance based on $s := 10,000$ simulated sample paths using the above algorithm, i.e.,

$$E(U) \approx \frac{1}{s} \sum_{k=1}^{s} U(\omega_k), \quad \text{Var}(U) \approx \frac{1}{s} \sum_{k=1}^{s} \left(U(\omega_k) - \frac{1}{s} \sum_{j=1}^{s} U(\omega_j) \right)^2.$$

According to the numerical results (see Figs. 1–3), the approximate probability density functions are well-behaved for both numerical methods. The means and

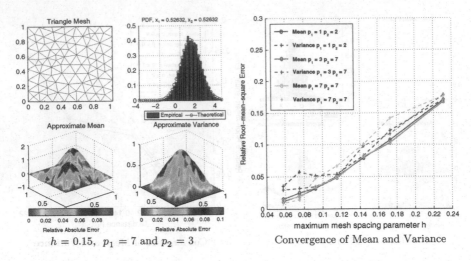

$h = 0.15, \ p_1 = 7 \text{ and } p_2 = 3$

Convergence of Mean and Variance

Fig. 3 Galerkin finite element methods, $n = 2$

variances of the kernel-based collocation solutions are smooth estimators, while the means and variances of the finite element solutions are piecewise smooth estimators. If we suitably truncate the Gaussian noise, then the finite element method has the same convergence rate as the kernel-based collocation method. If not, the kernel-based collocation method seems to do better than the finite element method for the variance.

Appendix A. Reproducing Kernels and Gaussian Fields

Let $K : \mathscr{D} \times \mathscr{D} \to \mathbb{R}$ be a reproducing kernel and $H_K(\mathscr{D})$ be its reproducing-kernel Hilbert space (see [10, Definition 10.1]). If $K \in C(\overline{\mathscr{D}} \times \overline{\mathscr{D}})$ then its *integral-type* kernel $\overset{*}{K}$ is defined by

$$\overset{*}{K}(x, y) := \int_{\mathscr{D}} K(x, z) K(y, z) \mathrm{d}z, \quad x, y \in \mathscr{D}. \tag{5}$$

Remark 4. As in the proof of [4, Lemma 2.2], in order to "match the spaces", any other kernel that "dominates" K (in the sense of [8]) could play the role of the integral-type kernel $\overset{*}{K}$.

Let $\mathscr{H}^m(\mathscr{D})$ be the *classical L_2-based Sobolev space* of order $m \in \mathbb{N}$. The *differential operator* $P : \mathscr{H}^m(\mathscr{D}) \to L_2(\mathscr{D})$ and the *boundary operator* $B : \mathscr{H}^m(\mathscr{D}) \to L_2(\partial\mathscr{D})$ are linear combinations of the derivatives D^α, $\alpha \in \mathbb{N}_0^d$, with nonconstant coefficients defined on $\mathscr{D}$ and $\partial\mathscr{D}$ respectively, i.e., $P = \sum_{|\alpha| \le m} c_\alpha D^\alpha$

and $B = \sum_{|\alpha| \leq m-1} b_\alpha D^\alpha|_{\partial \mathscr{D}}$ where $c_\alpha \in C(\mathscr{D})$ and $b_\alpha \in C(\partial \mathscr{D})$. Their orders are given by $\mathscr{O}(P) := m$ and $\mathscr{O}(B) := m - 1$.

Definition 1 ([3, Definition 3.28]). A stochastic process $S : \mathscr{D} \times \Omega \rightarrow \mathbb{R}$ is said to be *Gaussian* with mean $\mu : \mathscr{D} \rightarrow \mathbb{R}$ and covariance kernel $\Phi : \mathscr{D} \times \mathscr{D} \rightarrow \mathbb{R}$ on a probability space $(\Omega, \mathscr{F}, \mathbb{P})$ if, for any pairwise distinct points $X_\mathscr{D} := \{x_1, \cdots, x_N\} \subset \mathscr{D}$, the random vector $S := (S_{x_1}, \cdots, S_{x_N})^T$ is a multi-normal random variable on $(\Omega, \mathscr{F}, \mathbb{P})$ with mean μ and covariance matrix Φ, i.e., $S \sim \mathcal{N}(\mu, \Phi)$, where $\mu := (\mu(x_1), \cdots, \mu(x_N))^T$ and $\Phi := (\Phi(x_j, x_k))_{j,k=1}^{N,N}$.

We view the reproducing-kernel Hilbert space $H_K(\mathscr{D})$ as a *sample space* and its Borel σ-field $\mathscr{B}(H_K(\mathscr{D}))$ as a σ-*algebra* to set up the probability spaces.

Theorem 1 ([4, Theorem 3.1]). *Suppose that the reproducing kernel Hilbert space $H_K(\mathscr{D})$ is embedded into the Sobolev space $\mathscr{H}^m(\mathscr{D})$ with $m > d/2$. Further assume that the differential operator P and the boundary operator B have the orders $\mathscr{O}(P) < m - d/2$ and $\mathscr{O}(B) < m - d/2$. Given a function $\mu \in H_K(\mathscr{D})$ there exists a probability measure $\mathbb{P}^\mu$ defined on $(\Omega_K, \mathscr{F}_K) = (H_K(\mathscr{D}), \mathscr{B}(H_K(\mathscr{D})))$ such that the stochastic processes PS, BS given by*

$$PS_x(\omega) = PS(x, \omega) := (P\omega)(x), \quad x \in \mathscr{D} \subset \mathbb{R}^d, \quad \omega \in \Omega_K = H_K(\mathscr{D}),$$

$$BS_x(\omega) = BS(x, \omega) := (B\omega)(x), \quad x \in \partial \mathscr{D}, \quad \omega \in \Omega_K = H_K(\mathscr{D}),$$

are Gaussian fields with means $P\mu$, $B\mu$ and covariance kernels $P_1 P_2 \overset{*}{K}$, $B_1 B_2 \overset{*}{K}$ defined on $(\Omega_K, \mathscr{F}_K, \mathbb{P}^\mu)$, respectively, where

$$P_1 P_2 \overset{*}{K}(x, y) :- P_{z_1} P_{z_2} \overset{*}{K}(z_1, z_2)|_{z_1=x, z_2=y}, \quad x, y \in \mathscr{D},$$

$$B_1 B_2 \overset{*}{K}(x, y) := B_{z_1} B_{z_2} \overset{*}{K}(z_1, z_2)|_{z_1=x, z_2=y}, \quad x, y \in \partial \mathscr{D}.$$

When $P := I$ then we rewrite $PS = S$ which indicates that $S_x(\omega) = \omega(x)$.

Remark 5. The probability measure $\mathbb{P}^\mu$ defined in Theorem 1 can be verified to be *Gaussian* (see [3, 4, 11]). The papers [6, 7, 11] show what kinds of reproducing kernels satisfy the conditions that their reproducing-kernel Hilbert spaces are embedded into the Sobolev spaces. One typical example is the Sobolev spline kernel computed by the Matérn function.

Given $X_\mathscr{D} := \{x_j\}_{j=1}^N \subset \mathscr{D}$ and $X_{\partial \mathscr{D}} := \{x_{N+j}\}_{j=1}^M \subset \partial \mathscr{D}$, [4, Corollary 3.2] shows that the random vector $S_{PB} := (PS_{x_1}, \cdots, BS_{x_{N+M}})^T$ defined on $(\Omega_K, \mathscr{F}_K, \mathbb{P}^\mu)$ has a *multi-normal* distribution with mean m_{PB}^μ and covariance matrix $\overset{*}{K}_{PB}$, i.e.,

$$S_{PB} \sim \mathcal{N}(m_{PB}^\mu, \overset{*}{K}_{PB}),$$

where $m_{PB}^\mu := (P\mu(x_1), \cdots, P\mu(x_N), B\mu(x_{N+1}), \cdots, B\mu(x_{N+M}))^T$ and

$$\overset{*}{\mathsf{K}}_{PB} := \begin{pmatrix} (P_1 P_2 \overset{*}{K}(x_j, x_k))_{j,k=1}^{N,N}, & (P_1 B_2 \overset{*}{K}(x_j, x_{N+k}))_{j,k=1}^{N,M} \\ (B_1 P_2 \overset{*}{K}(x_{N+j}, x_k))_{j,k=1}^{M,N}, & (B_1 B_2 \overset{*}{K}(x_{N+j}, x_{N+k}))_{j,k=1}^{M,M} \end{pmatrix}.$$

Fix any $x \in \mathscr{D}$. We can compute the joint probability density function p_X^μ and p_J^μ of S_{PB} and (S_x, S_{PB}) defined on $(\Omega_K, \mathscr{F}_K, \mathbb{P}^\mu)$ respectively. By Bayes' rule, we can obtain the *conditional probability density function* p_x^μ of S_x given S_{PB} defined on $(\Omega_K, \mathscr{F}_K, \mathbb{P}^\mu)$, i.e., for each $v \in \mathbb{R}^{N+M}$,

$$p_x^\mu(v|v) := \frac{p_J^\mu(v, v)}{p_X^\mu(v)} = \frac{1}{\sigma(x)\sqrt{2\pi}} \exp\left(-\frac{(v - m_x^\mu(v))^2}{2\sigma(x)^2} \right), \quad v \in \mathbb{R}, \quad (6)$$

where $m_x^\mu(v) := \mu(x) + k_{PB}(x)^T \overset{*}{\mathsf{K}}_{PB}{}^\dagger (v - m_{PB}^\mu)$, $\sigma(x)^2 := \overset{*}{K}(x,x) - k_{PB}(x)^T \overset{*}{\mathsf{K}}_{PB}{}^\dagger k_{PB}(x)$ and $k_{PB}(x) := (P_2 \overset{*}{K}(x, x_1), \cdots, B_2 \overset{*}{K}(x, x_{N+M}))^T$. In particular, given the real observation $y := (y_1, \cdots, y_{N+M})^T$, S_x conditioned on $S_{PB} = y$ has the probability density $p_x^\mu(\cdot|y)$.

References

1. I. Babuška, F. Nobile, R. Tempone, A stochastic collocation method for elliptic partial differential equations with random input data. SIAM Rev. **52**, 317–355 (2010)
2. I. Babuška, R. Tempone, G.E. Zouraris, Galerkin finite element approximations of stochastic elliptic partial differential equations. SIAM J. Numer. Anal. **42**, 800–825 (2004)
3. A. Berlinet, C. Thomas-Agnan, *Reproducing Kernel Hilbert Spaces in Probability and Statistics* (Kluwer, Boston, 2004)
4. I. Cialenco, G.E. Fasshauer, Q. Ye, Approximation of stochastic partial differential equations by a kernel-based collocation method. Int. J. Comput. Math. Special issue: Recent advances on the numerical solutions of stochastic partial differential equations. (2012), doi:10.1080/00207160.2012.688111
5. G.E. Fasshauer, *Meshfree Approximation Methods with Matlab* (World Scientific, Singapore, 2007)
6. G.E. Fasshauer, Q. Ye, Reproducing kernels of generalized Sobolev spaces via a Green function approach with distributional operators. Numer. Math. **119**, 585–611 (2011)
7. G.E. Fasshauer, Q. Ye, Reproducing kernels of Sobolev spaces via a Green kernel approach with differential operators and boundary operators. Adv. Comput. Math., to appear, doi:10.1007/s10444-011-9264-6.
8. M.N. Lukić, J.H. Beder, Stochastic processes with sample paths in reproducing kernel Hilbert spaces. Trans. Amer. Math. Soc. **353**, 3945–3969 (2001)
9. M. Scheuerer, R. Schaback, M. Schlather, *Interpolation of Spatial Data – A Stochastic or A Deterministic Problem?*, submitted data page of R. Schaback's Research Group, http://num.math.uni-goettingen.de/schaback/research/group.html
10. H. Wendland, *Scattered Data Approximation* (Cambridge University Press, Cambridge, 2005)
11. Q. Ye, Analyzing reproducing kernel approximation methods via a Green function approach, Ph.D. thesis, Illinois Institute of Technology, 2012

A Meshfree Method for the Analysis of Planar Flows of Inviscid Fluids

Vasily N. Govorukhin

Abstract A variant of a vortex particle-in-cell method is proposed for computing 2D inviscid incompressible flows in a closed domain. The governing equations are the 2D Euler equations in terms of the stream function and the vorticity, or geophysical models of the atmosphere. The approach is based on an approximation of the vorticity field using its values at a set of fluid particles and the stream function is computed by a Galerkin method. The flow domain is divided into rectangular cells. Vorticity in every cell is interpolated by a third order polynomial. The resultant piecewise continuous polynomial approximation of the vorticity is employed to derive analytically Galerkin's coefficients of the stream function expansion. A parallel algorithm of the proposed method is also presented. The performance of the method is illustrated by some numerical results dealing with flows trough channels and the dynamics of the multipoles vortex patch calculation.

Keywords Incompressible inviscid flows • Vortices in cells method • Vortex dynamics

1 Introduction

In many situations, an inviscid incompressible flow is the simplest appropriate model of a real fluid dynamics processes. A numerical analysis based on the Eulerian or Lagrangian approach is widely used to address flow problems. Examples are various Lagrangian vortex methods (see [4] and the references therein), which are based on solving equations determining the vorticity distribution in

V.N. Govorukhin (✉)
Southern Federal University, Milchakova str. 8a, Rostov-on-Don, Russia
e-mail: vgov@math.sfedu.ru

M. Griebel and M.A. Schweitzer (eds.), *Meshfree Methods for Partial Differential Equations VI*, Lecture Notes in Computational Science and Engineering 89, DOI 10.1007/978-3-642-32979-1_11, © Springer-Verlag Berlin Heidelberg 2013

the flow field. Given a vorticity field in N_p fluid particles, one can calculate the velocities of the particles, while the vorticity itself is transported with the particles. Convergence analysis and error estimates are available for vortex methods (see [3, 4, 8]). These methods are especially popular if the flow region is the entire plane, since the velocity is then explicitly expressed from the vorticity field in terms of the Green's function of Poisson's equation. In the case of a closed flow region in the presence of additional physical factors, Poisson's equation has to be solved numerically. In this case, vortex methods can be accelerated using combined Eulerian–Lagrangian vortex techniques. In this paper, a vortex method is developed for analyzing unsteady inviscid flows in a vessel with permeable boundaries and with some geophysical assumptions. The first version of this method was described in [5] and extended in [7]. In [6] the method was applied for analysis of flows through rectangular channels.

2   Governing Equations

The dynamics of an incompressible inviscid fluid is described by the Euler equation

$$\frac{D\omega}{Dt} \equiv \omega_t + \psi_y \omega_x - \psi_x \omega_y = 0, \tag{1}$$

where ω denotes the vorticity, ψ is the stream function and D/Dt denotes the material derivative. Here $\psi_x = \partial\psi/\partial x$, $\psi_y = \partial\psi/\partial y$, $\psi_{xx} = \partial^2\psi/\partial x^2$, etc. The velocity of the fluid $v = (v_1, v_2)$ is expressed via the stream function ψ as

$$v_1 = \psi_y, \quad v_2 = -\psi_x, \tag{2}$$

Equation (1) means that ω is passively transported by fluid particles. In the most general case of two dimensional barotropic fluid dynamics on a sphere with allowance for the Coriolis force, the preserved quantity is the potential vorticity, which is related to the stream function as follows

$$\omega = -\Delta\psi + \Lambda^2\psi - \frac{1}{2}\gamma r^2. \tag{3}$$

Here, the Coriolis parameter near a pole is given by $f(r) = f_0 - \frac{1}{2}\gamma r^2 + O(r^4)$, where $\gamma = const$, $r = \sqrt{x^2 + y^2}$ is the polar radius, $\Lambda^2 = f_0^2/gh = const$, g is the acceleration due to gravity, and h is the thickness of the fluid layer.

The initial condition for the vorticity is specified as

$$\omega|_{t=0} = \omega_0(x, y) \tag{4}$$

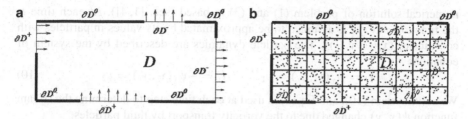

Fig. 1 The computational domain (**a**) and an example of a cell distribution (**b**) with particles used to approximate function $\omega(x, y)$

Problems (1), (3) and (4) can be considered in various domains: in the entire plane, a half-plane, or a closed vessel with various boundary conditions (see [9, 12, 15, 19]). Here, we study the dynamics of a fluid in a closed domain D with a possible flow through it. In this case, the flow velocity on the domain boundary ∂D is set as

$$\psi|_{\partial D} = \psi^{\partial D}. \tag{5}$$

The boundary of D is divided into three portions (see Fig. 1a): ∂D^+ (the normal flow velocity is directed inward into D), ∂D_0 (the normal velocity is zero), and ∂D^- (the normal velocity is directed outward from inside D) with the net flux through the boundary being zero. The boundary condition for ω is set only on ∂D^+:

$$\omega|_{\partial D^+} = \omega^{\partial D^+}. \tag{6}$$

For computational convenience, we make the substitution

$$\psi = \Psi + \psi^{\partial D} \tag{7}$$

Then Eq. (3) becomes

$$-\Delta\Psi + \Lambda^2\Psi = \omega + \Delta\psi^{\partial D} - \Lambda^2\psi^{\partial D} + \frac{1}{2}\gamma r^2, \tag{8}$$

and the boundary conditions for Ψ are

$$\Psi|_{\partial D} = 0. \tag{9}$$

3 Numerical Method

It follows from Eq. (1) that the vorticity is passively transported by the velocity field (v_1, v_2); i.e., the initial vorticity value in a fluid particle is preserved in time: $\omega(x_i(t), y_i(t)) = \omega_0(x_i(t_0), y_i(t_0))$. On this property the vortex method for the numerical solution of problem (1) and (3) is based (see, [1, 4]). At each time t, the function $\omega(x, y)$ is assumed to be approximated by its values in particles with coordinates $(x_i(t), y_i(t))$. The particle dynamics are described by the system of equations

$$\dot{x}_i = \psi_y(x_i, y_i) = v_1, \qquad \dot{y}_i = -\psi_x(x_i, y_i) = v_2 \tag{10}$$

We solve system (10), and Eq. (3) is used at each time step to determine the stream function $\psi(x, y)$ changed due to the vorticity transport by fluid particles.

Vortex methods differ from each other primarily in the approach used to solve Eq. (3). They involve an explicit expression for the stream function $\psi(x, y) = \int_D G\omega(x, y)dxdy$, where G is the corresponding Green operator (for example for the case $\Lambda = \gamma = 0$ and D is entire plane). In this case, we need an explicit expression for the Green operator and use quadrature formulas (see [11, 17]). In other variants of vortex methods, Eq. (3) is solved numerically, for example, by finite differences [14] or the projection method [5]. Due to this approach, vortex methods can be used in various flow regions; additionally, the computational costs can be reduced and a larger number of vortex elements can be used.

After making substitution (7), the particles dynamic equations (10) become

$$\dot{x}_i = \Psi_y(x_i, y_i) + \psi_y^{\partial D}, \qquad \dot{y}_i = -\Psi_x(x_i, y_i) - \psi_x^{\partial D} \tag{11}$$

Initially, N_p particles with coordinates $(x_i, y_i), i = 1, \ldots, N_p$, and with the vorticity $\omega_i = \omega_0(x_i, y_i)$ determined by initial condition (4) are given in D. The particles can be defined deterministically or randomly, and their initial distribution does not need to be uniform. In the case of a flow through the domain, a particle may escape the channel in the course of the computations, in which case a new particle with the vorticity determined by boundary condition (6) is specified at the inlet.

To solve system (11), the function $\Psi(x, y)$ has to be found at each point of D at the time t.

3.1 Approximation of the Velocity Field

Further the case of rectangular area of a flow D is considered:

$$D = \left\{(x, y) : -\frac{a}{2} \leq x \leq \frac{a}{2}; -\frac{b}{2} \leq y \leq \frac{b}{2}\right\}. \tag{12}$$

To solve the problem (8) and (9) for each t, we employ the Bubnov-Galerkin method with the basis functions

$$g_{ij}(x, y) = \sin\left(\frac{i\pi}{a}\left(x + \frac{a}{2}\right)\right) \sin\left(\frac{j\pi}{b}\left(y + \frac{b}{2}\right)\right) \tag{13}$$

We seek a solution in the form

$$\Psi \approx \widetilde{\Psi} = \sum_{i=1}^{k_x} \sum_{j=1}^{k_y} \Psi_{i,j} g_{ij}(x, y) \tag{14}$$

Substituting (14) into Eq. (14) and applying the operations of projection onto the basis functions, we obtain the unknown coefficients $\Psi_{i,j}$:

$$\Psi_{i,j} = \left[\frac{\gamma ab \left(i^2 j^2 \pi^2 (a^2+b^2) - 8(j^2 a^2 + i^2 b^2) \right) \left((-1)^i - 1 \right) \left((-1)^j - 1 \right)}{8\pi^4 i^3 j^3} \right. $$
$$+ \int_{-a/2}^{a/2} \int_{-b/2}^{b/2} \left(\Delta \psi^{\partial D} - \Lambda^2 \psi^{\partial D} \right) g_{i,j}(x, y) \, dx dy \tag{15}$$
$$\left. + \int_{-a/2}^{a/2} \int_{-b/2}^{b/2} \omega(x, y) \, g_{i,j} \, dx dy \right] \times \frac{4ab}{\pi^2 (i^2 b^2 + j^2 a^2) + \Lambda^2 a^2 b^2}$$

The multiplier outside the square brackets rapidly tends to zero with increasing i and j. As a result, we can use a bounded number of terms in sum (14). The first integral in (15) is evaluated analytically or numerically once in the beginning of calculations. Thus, the determination of expansion coefficients (15) is reduced to the evaluation of the integrals

$$I_{i,j} = \int_{-a/2}^{a/2} \int_{-b/2}^{b/2} \omega(x, y) \, g_{i,j} \, dx dy \tag{16}$$

For approximate $\omega(x, y)$, the domain D is divided into N_{box} rectangular cells, see Fig. 1b, where $N_{box} = n_x \times n_y$. In each cell k, the function $\omega(x, y)$ is approximated by a polynomial in two variables:

$$\omega(x, y) \approx \sum_{k=1}^{N_{box}} \phi_k(x, y) = \sum_{k=1}^{N_{box}} \sum_{l,m=0, l+m \leq 3}^{3} a_{k,l,m} x^l y^m. \tag{17}$$

In each cell, the coefficients $a_{k,l,m}$ in (17) are found using the least squares method by minimizing the expression

$$S_k = \sum_n \left(\sum_{l,m=0, l+m \leq 3}^{3} a_{k,l,m} x_n^l y_n^m - \omega(x_n, y_n) \right)^2 \tag{18}$$

The computational complexity of determining all $a_{k,i,j}$ is proportional to $N_box x$, which is not expensive. After $a_{k,i,j}$ were found, integrals (16) are calculated as

$$I_{ij} \approx \sum_{k=1}^{N_{box}} \sum_{l,m=0, l+m \leq 3}^{3} \int_{-a/2}^{a/2} \int_{-b/2}^{b/2} a_{k,l,m} x^l y^m g_{ij}(x, y) dx dy. \tag{19}$$

In the case of basis functions (13), the integrals in (19) are easily evaluated analytically, which considerably reduces the amount of computations.

3.2 Particle Dynamics Computation

After finding coefficients $\Psi_{i,j}$, the discrete velocity field of the fluid particle is given by

$$\tilde{v}_1 = \dot{x}_i = \sum_{i=1}^{k_x} \sum_{j=1}^{k_y} \Psi_{i,j} \frac{\partial g_{ij}}{\partial y} + \psi_y^{\partial D}, \quad \tilde{v}_2 = \dot{y}_i = -\sum_{i=1}^{k_x} \sum_{j=1}^{k_y} \Psi_{i,j} \frac{\partial g_{ij}}{\partial x} - \psi_x^{\partial D} \quad (20)$$

System (20) is Hamiltonian with the Hamiltonian $\tilde{\Psi} + \psi^{\partial D}$. A fundamental properties of such systems is that they preserve the phase volume in time, or are symplectic. It is well known (see [16]) that conservation of this properties is very important for numerical solution. Symplecticity preserving integrators are implicit, which complicates their application in the case of high order systems. A possible way of overcoming this difficulty is to use explicit-implicit predictor-corrector schemes with an explicit predictor and a symplectic corrector (see, for example [18]). In [2] explicit pseudo-symplectic methods were proposed that preserve symplecticity to higher order accuracy than the accuracy of the numerical solution. Efficiency of these methods for calculation of dynamics of liquid particles on the long times was shown experimentally in [5,7]. Pseudo-symplectic methods are used in the presented numerical scheme for time integration of the system (20).

3.3 Parallel Algorithm

For the parallelization, the numerical method was modified using the OpenMP library for parallel computers. Most of the computational cost of the flow calculation at each time-step corresponds to the finding stream function as a solution of Eq. (8). The integration of equations (20) was realized without parallelization with using explicit pseudo-symplectic methods PS36 with time step h (see, [2]). The algorithm of parallel version of method is summarized as follows:

For each value of t.
1. **do in parallel, i = 1:N_p:** *Sorting particles by cells.*
2. **do in parallel, i = 1:N_{box}:** *Finding coefficients $a_{k,l,m}$ as a solution of (18) for each cell.*
3. **do in parallel, i = 1:N_{box}:** *Calculation $\Psi_{i,j}$ using (15).*
4. **do in parallel, i = 1:N_p:** *Calculation the velocity $(\tilde{v}_1, \tilde{v}_2)$ for each particle.*

Several tests up to eight CPUs have been performed to estimate performance of the parallel algorithm. The results of five calculations with different parameters of methods are presented in Fig. 2. The more effective results (more fivefold acceleration for 8 CPUs) was obtained for more detailed approximation ($N_p = 240,000$, $N_{box} = 60 \times 20 = 1,200$, $k_x = k_y = 60$).

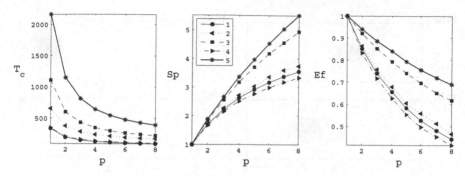

Fig. 2 Time of calculation (*left*), speedup (*center*) and efficiency (*right*) of the parallel algorithm with respect to the number of processors and for five sets of domain sizes and parameters of method

4 Numerical Experiments

An effective method for testing a numerical scheme is the reproduction of analytical facts or results obtained apolitically or by other computational methods. We use three example of evolution of known vortex structures as for test calculations. All the computations are made with $\Lambda = \gamma = 0$.

Example 1. We carry out a calculations of the evolution of an initially elliptic vortex patch in the presence of inflow and outflow through the boundary of the domain. Numerical experiments was provided for duct $D : 0 \leq x \leq 3, 0 \leq y \leq 1$. The following boundary and initial conditions was used:

$$\psi^{\partial D} = Q_1 y, \quad \omega^{\partial D} = 0, \quad \omega_0(x, y) = e^{-45(x-3/4)^2 - 15(y-1/2)^2} \tag{21}$$

Figure 3 demonstrate the dynamics of elliptical vortex patch for $Q_1 = 0.04$. Results well coincide with calculations for all plane, see [11].

Example 2. Another test was computed in a rectangle D with the same lengths under the boundary conditions

$$\psi^{\partial D}|_{x=\pm a/2} = 0.1 \left(y + \tfrac{b}{2}\right), \quad \psi^{\partial D}|_{y=-b/2} = 0,$$
$$\psi^{\partial D}|_{y=b/2} = 0.1b, \quad \omega^{\partial D+}|_{x=-a/2} = 0,$$

with the initial vorticity distribution (see [13])

$$\omega_0 = 40 \left(0.5 - y\right) e^{-100\left(y-\frac{1}{2}\right)^2} \left(\tanh(50x - 25) - \tanh(50x - 125) \right). \tag{22}$$

which is shown in Fig. 4 at t = 0. This test was computed on the time interval $t \in [0, 10]$ with $h = 0.02$, $kx = ky = 30$, $nx = 60$, $ny = 20$, and $Np = 100,000$. The result was the vortex configuration depicted in figure at t = 10, which qualitatively agrees with its counterpart in [13]. As in [13], a dipole was formed from initial field (22).

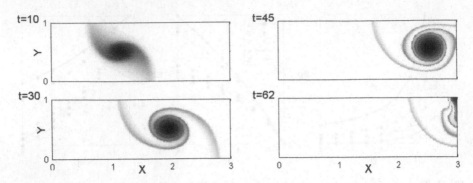

Fig. 3 Vorticity distribution for different t and uniform flow with velocity $Q_1 = 0 : 04$ on the inlet and outlet of D. White color corresponds to minimal vorticity (zero) and black to maximal (one)

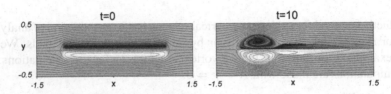

Fig. 4 A dipole formation from initial field (22)

Example 3. A family of semianalytical solutions describing multipolar vortical structures with zero total circulation in a variety of two-dimensional models was presented in [10]. A multipole, termed also an $m + 1$-pole, is a vortical system that possesses an m-fold symmetry and is comprised of a central core vortex and m satellite vortices surrounding the core. Fluid parcels in the core and the satellites revolve oppositely, and the multipole as a whole rotates steadily. In time $t = 2\pi$ the vortex configuration does one complete turn.

Numerical experiments was provided for duct $D : -5 \le x \le 5, -5 \le y \le 5$. The boundary and initial conditions are defined by the analytical solutions, see [10]. This test was computed on the time interval $t \in [0, 20\pi]$ with $h = \pi/50$, $kx = ky = 36$, $nx = 60, ny = 60$, and $N_p = 120,000$. In Fig. 5 results of calculation of a quadrupole vortical configuration are presented. It is visible that for ten complete turns about the center the configuration practically did not change. It shows efficiency of a described method for calculation of dynamics on enough big times.

The presented results of three test calculations completely reproduce the known results. Calculations demanded insignificant computer resources. It shows efficiency of the presented computational method of dynamics calculation of vortical structures.

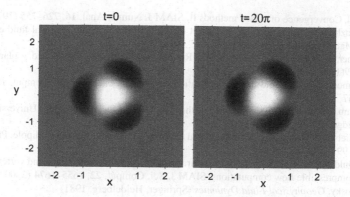

Fig. 5 Vortex field for quadrupole configuration on the start $t = 0$ and after 10 turns $t = 2\pi$.

5 Concluding Remarks

We presented a general approach to the treatment of inviscid fluid flows in two space dimensions in this paper. We have implemented the proposed scheme and presented some numerical results which clearly demonstrate the viability of our approach. A numerical method was proposed that applies to a wide range of fluid dynamic problems in two dimensions. An advantage of the method is its relatively low computational cost, which makes it possible to compute vortex structures at sufficiently long times with a high degree of detail. The simplest application of the method to an inviscid flow in a rectangular region is discussed in this paper, but this approach can be extended and adapted to other domains and viscous flows.

Acknowledgements The author would like to acknowledge the grants 11-01-00708 and 12-01-00668 from the Russian Foundation for Basic Research.

References

1. C. Anderson, C. Greengard, On vortex methods. SIAM J. Numer. Anal. **22**, 413–440 (1985)
2. A. Aubry, P. Chartier, Pseudo-symplectic runge-kutta methods. BIT **38**, 439–461 (1998)
3. J.T. Beale, A. Majda, Vortex methods: II. Higher order accuracy in two and three dimensions. Math. Comput. **39**, 29–52 (1982)
4. G.H. Cottet, P.D. Koumoutsakos, *Vortex Methods: Theory and Practice* (Cambridge University Press, Cambridge, 1999)
5. V.N. Govorukhin, K.I. Ilin, Numerical study of an inviscid incompressible flow through a channel of finite length. Int. J. Numer. Method Fluid **60**, 1315–1333 (2009)
6. V.N. Govorukhin, A.B. Morgulis, V.A. Vladimirov, Planar inviscid flows in a channel of finite length: washout, trapping and self-oscillations of vorticity. J. Fluid Mech. **659**, 420–472 (2010)
7. V.N. Govorukhin, A vortex method for computing two-dimensional inviscid incompressible flows. Comput. Math. Math. Phys. **51**(6), 1061–1073 (2011)

8. O. Hald, Convergence of vortex methods II. SIAM J. Numer. Anal. **16**, 726–755 (1979)
9. A.V. Kazhikhov, Remark on the statement of the flow problem for the ideal fluid equations. Prikl. Mat. Mekh. **44**, 947–949 (1980)
10. Z. Kizner, R. Khvoles, J.C. McWilliams, Rotating multipoles on the f- and γ-planes. Phys. Fluid **19**(1), 016603, 13 (2007)
11. P. Koumoutsakos, Inviscid axisymmetrization of an elliptical vortex. J. Comput. Phys. **138**, 821–857
12. H. Lamb, *Hydrodynamics*. Reprint of the 1932 6th edn. (Cambridge University Press, Cambridge, 1993)
13. A.H. Nielsen, J. Rasmussen, Formation and temporal evolution of the lamb-dipole. Phys. Fluid **9**, 982–991 (1997)
14. M.L. Ould-Salihi, G.-H. Cottet, M. El Hamraoui, Blending finite difference and vortex methods for incompressible flow computations. SIAM J. Sci. Comput. **22**, 1655–1674 (2000)
15. J. Pedlosky, *Geophysical Fluid Dynamics* (Springer, Heidelberg, 1981)
16. J.M. Sanz-Serna, M.P. Calvo, *Numerical Hamiltonian Problems* (Chapman and Hall, London, 1994)
17. J. Strain, 2D vortex methods and singular quadrature rules. J. Comput. Phys. **124**, 131–145 (1996)
18. P.W.C. Vosbeek, R.M.M. Mattheij, Contour dynamics with symplectic time integration. J. Comput. Phys. **133**, 222–234 (1997)
19. V.I. Yudovich, A two-dimensional nonstationary problem of the flow of an ideal incompressible fluid through a given region. Mat. Sb. **64**, 562–588 (1964)

Some Regularized Versions of the Method of Fundamental Solutions

Csaba Gáspár

Abstract A powerful method of the solution of homogeneous equations is considered. Using the traditional approach of the Method of Fundamental Solutions, the fundamental solution has to be shifted to external source points. This is inconvenient from computational point of view, moreover, the resulting linear system can easily become severely ill-conditioned. To overcome this difficulty, a special regularization technique is applied. In this approach, the original second-order elliptic problem (a modified Helmholtz problem in the paper) is approximated by a fourth-order multi-elliptic boundary interpolation problem. To perform this boundary interpolation, either the Method of Fundamental Solutions, or a direct multi-elliptic interpolation can be used. In the paper, a priori error estimations are deduced. A numerical example is also presented.

Keywords Method of fundamental solutions • Multi-elliptic interpolation • Regularization • Radial basis functions • Quadtree • Multigrid

1 Introduction

The Method of Fundamental Solutions (MFS) is a truly meshfree method, which requires neither domain nor boundary element structure [1]. The approximate solution is a linear combination of the fundamental solution of the elliptic equation to be solved shifted to some source points. Due to the singularity of the fundamental solution of the second-order elliptic operators, the source points should be located *outside* the domain, which is often inconvenient from computational point of view. The choice of the location of the source points is crucial. If they are too far from the

C. Gáspár (✉)
Széchenyi István University, P.O. Box 701, H-9007 Győr, Hungary
e-mail: gasparcs@sze.hu

M. Griebel and M.A. Schweitzer (eds.), *Meshfree Methods for Partial Differential Equations VI*, Lecture Notes in Computational Science and Engineering 89, DOI 10.1007/978-3-642-32979-1_12, © Springer-Verlag Berlin Heidelberg 2013

boundary, the approximate linear system becomes highly ill-conditioned. If they are too close to the boundary, numerical singularities appear.

To avoid the problem of the singularities, special techniques have been developed. In the so-called desingularization methods, the appearing singular matrix entries are computed in a special way, thus making it possible to locate the source points on the boundary [4, 10, 11]. The boundary knot method [3] uses nonsingular general solutions as radial basis functions instead of fundamental solutions, which allows the source and collocation points (in which the boundary condition is enforced) to coincide. Unfortunately, the resulting linear system is often even more ill-conditioned than in the case of the MFS. More recently, a regularized method of fundamental solutions (RMFS) has been proposed [7], where the original second-order problem is approximated by a singularly perturbed fourth-order one. However, the fundamental solution of the applied fourth-order operator is continuous at the origin, thus, the new problem can be approximately solved by the MFS without difficulty. The source points can be located along the boundary. This significantly reduces the condition number of the resulting linear system at the same time. On the other hand, however, the exactness of this technique remains below that of the traditional MFS. In [7], error estimations have been derived for the approximate fourth-order problem.

In this paper, error estimations will be deduced for the regularized MFS-solution itself. These estimations are based on the multi-elliptic interpolation [5]. In this context, the RMFS can be regarded as an RBF-type boundary interpolation method, which uses the fundamental solution of the above fourth-order operator as a radial basis function. It should be pointed out that this boundary interpolation can be performed *directly* by solving the fourth-order problem using quadtrees and multigrid tools as proposed in [5, 6]. This makes it possible to completely avoid the use of large dense and ill-conditioned matrices.

2   The Method of Fundamental Solutions

As a model example, consider the 2D modified Helmholtz equation supplied with Dirichlet boundary condition:

$$\Delta u - \lambda^2 u = 0 \quad \text{in } \Omega, \quad u|_\Gamma = u_0 \tag{1}$$

Here $\Omega \subset \mathbb{R}^2$ is a smooth domain with boundary Γ, and $\lambda \geq 0$ is the Helmholtz constant. Introduce a finite set of *source points* $\tilde{x}_1, \tilde{x}_2, \ldots, \tilde{x}_N$ outside of Ω and another set of *collocation points* $x_1, x_2, \ldots, x_N$ along the boundary. The MFS defines an approximate solution of (1) in the form:

$$u(x) := \sum_{j=1}^{N} \alpha_j \Phi(x - \tilde{x}_j), \tag{2}$$

Table 1 Relative L_2-errors (%) of the MFS-solution of the modified Helmholtz test problem. N is the number of boundary collocation points, d is their distance from the boundary

$d \setminus N$	16	32	64	128	256
0.25	2.83	0.037	1.47E−5	4.59E−12	6.29E−14
0.50	0.523	3.56E−4	4.04E−10	7.03E−14	2.26E−13
1.00	0.103	7.32E−7	4.49E−14	5.49E−14	2.44E−13

Table 2 Condition numbers of the MFS-matrices, modified Helmholtz test problem. N is the number of boundary collocation points, d is their distance from the boundary

$d \setminus N$	16	32	64	128	256
0.25	7.568	84.50	5.85E+3	1.46E+7	4.63E+13
0.50	19.90	911.0	1.14E+6	9.62E+11	***
1.00	80.66	3.27E+4	3.85E+9	***	***

where $\Phi(x) = -\frac{1}{2\pi}K_0(\lambda\|x\|)$, the fundamental solution of (1) (K_0 denotes the usual modified Bessel function). The a priori unknown coefficients α_j are to be determined from the boundary condition, by solving the linear system:

$$\sum_{j=1}^{N}\alpha_j\,\Phi(x_k - \tilde{x}_j) = u_0(x_k) \quad (k = 1, 2, \ldots, N) \tag{3}$$

Since the Bessel function K_0 has a logarithmic singularity at the origin, the source points have to be located outside of the domain. The matrix of (3) is fully populated, nonsymmetric and severely ill-conditioned, if the source points are far from the boundary. On the other hand, if they are too close to the boundary, numerical singularities appear in the vicinity of the collocation points. These numerical disadvantages can be significantly reduced the regularized version of the MFS which is outlined in the next section.

To illustrate the above phenomena, consider the problem (1) with the exact test solution

$$u^*(x, y) := (\sin x) \cdot \sinh(\sqrt{1 + \lambda^2} \cdot y), \tag{4}$$

where we use the more familiar notations x, y for the spatial variables. In this test example, set $\lambda := 2$. Let the domain Ω of the problem be the unit circle. Eq. (1) is supplied with Dirichlet boundary condition in a consistent way. The source points are located at a distance d from the boundary of Ω. Both the source points and the collocation points are equally spaced along the circles having radii 1 and $1 + d$, respectively. Table 1 shows the relative L_2-errors of the MFS-solutions with different numbers of source points (N) and distances from the boundary (d), whereas Table 2 shows the corresponding condition numbers (the notation *** indicates that the computed condition number exceeds the value 1.0E+16). It can clearly be seen that the MFS results in very good approximate solutions, however, the condition numbers increase extremely rapidly, when the number of sources or their distance from the boundary increases.

3   The Regularized Method of Fundamental Solutions

Consider again the model problem (1) and the following fourth-order problem:

$$(\Delta - c^2 I)(\Delta - \lambda^2 I)u = 0 \quad \text{in } \Omega, \quad u|_\Gamma = u_0, \quad \frac{\partial u}{\partial n}|_\Gamma = v_0 \qquad (5)$$

with arbitrary (sufficiently regular) Neumann datum v_0 (I denotes the identity operator). The constant c plays a scaling role and needs a careful definition. In practice, $c \gg \lambda$ and should depend on the characteristic distance of the boundary collocation points.

It has been proved in [7] that – at least, when Ω is a half-stripe and $\lambda = 0$ – the solution u of (5) approximates well the solution u^* of (1). More precisely, the following estimation holds:

$$\|u - u^*\|_{L_2(\Omega)}^2 \leq \frac{C}{c^2}\left(\|u_0\|_{H^{1/2}(\Gamma)}^2 + \|v_0\|_{H^{-1/2}(\Gamma)}^2\right)$$

where the constant C is independent of the scaling constant c and also of the boundary data. That is, if c is large enough, the solution of (5) is a good approximation of the solution of the original problem, independently of the Neumann boundary condition in (5).

Remarks.

* Problem (5) can be regularized in another way as well, by replacing the Laplace operator with the fourth-order operator $\Delta(I - \frac{1}{c^2}\Delta)$:

$$\Delta\left(I - \frac{1}{c^2}\Delta\right)u - \lambda^2 u = 0$$

However, this equation is equivalent to $\Delta\Delta u - c^2 \Delta u + c^2 \lambda^2 u = 0$, while (1) is equivalent to a similar one: $\Delta\Delta u - (c^2 + \lambda^2)\Delta u + c^2 \lambda^2 u = 0$. For the sake of simplicity, we will use (5) in the following.
* If the original problem is a Laplace equation (i.e. $\lambda = 0$), then the fundamental solution of the operator $\Delta(\Delta - c^2 I)$ has to be used:

$$\Phi(x) = -\frac{1}{2\pi c^2}(K_0(c\|x\|) + \log(c\|x\|))$$

Note that this function is also continuous at the origin.
* The above technique fails when the original problem is supplied with Neumann or mixed boundary condition, since the normal derivative of the solution of the fourth-order problem (5) does not necessarily approximate the normal derivative of the solution of the original second-order problem (1). Therefore the presented RMFS-method in the above form is restricted to the case of the Dirichlet boundary condition.

To numerically solve the new problem (5), the MFS can be applied without difficulty. The main advantage of this approach is that the fundamental solution of the fourth-order operator $(\Delta - c^2 I)(\Delta - \lambda^2 I)$, which has the form:

$$\Phi(x) = -\frac{1}{2\pi} \cdot \frac{1}{c^2 - \lambda^2} (K_0(c\|x\|) - K_0(\lambda\|x\|))$$

is *continuous* at the origin (the singularities of the two terms cancel out). This makes it possible to allow the source and collocation points to coincide, which simplifies the use of the method in practice and also significantly reduces the condition number of the matrix appearing in (3).

Another numerical technique to solve the approximate problem (5) is the multi-elliptic boundary interpolation (see [8]). Here, instead of any MFS-like technique, the following idea is used. Let $\Omega_0 \supset \Omega$ be a larger domain and consider the following pair of problems:

Direct multi-elliptic interpolation: Find a function $v \in H_0^2(\Omega_0)$ which satisfies (5) with the exception of the boundary points $x_1, \ldots, x_N$ (in the sense of distributions). At these points, interpolation conditions are prescribed:

$$(\Delta - c^2 I)(\Delta - \lambda^2 I)v = 0 \quad \text{in } \Omega_0 \setminus \{x_1, \ldots, x_N\} \tag{6}$$

$$v(x_k) = u_0(x_k) \quad (k = 1, \ldots, N)$$

Variational multi-elliptic interpolation: Find a function $v \in H_0^2(\Omega_0)$ that minimizes the functional

$$F(v) := \|\Delta v\|_{L_2(\Omega)}^2 + (c^2 + \lambda^2)\|\text{grad } u\|_{L_2(\Omega)}^2 + c^2\lambda^2\|u\|_{L_2(\Omega)}^2 \tag{7}$$

among the functions of $H_0^2(\Omega_0)$ which satisfy the interpolation conditions $v(x_k) = u_0(x_k)$ $(k = 1, 2, \ldots, N)$.

The above two problems are equivalent and have a unique solution, which is a immediate generalization of the results of [6]. Furthermore, standard variational arguments show that if $u \in H_0^2(\Omega_0)$ is an arbitrary function satisfying the interpolation conditions, then $(u - v)$ is the orthogonal projection of u with respect to the scalar product

$$\langle w_1, w_2 \rangle := \langle \Delta w_1, \Delta w_2 \rangle_0 + (c^2 + \lambda^2)\langle \text{grad } w_1, \text{grad } w_2 \rangle_0 + c^2\lambda^2\langle w_1, w_2 \rangle_0$$

to the subspace

$$W := \{w \in H_0^2(\Omega_0) : w(x_k) = 0, \ k = 1, \ldots, N\},$$

which is closed in the Sobolev space $H_0^2(\Omega_0)$ (here $\langle ., . \rangle_0$ means the scalar product in $L_2(\Omega_0)$). For details, see [6].

The main advantage of the multi-elliptic interpolation is to make it possible
to completely avoid the use of large, dense and ill-conditioned matrices. Instead,
the multi-elliptic problem (6) has to be solved, preferably by using a quadtree cell
system (generated by the interpolation points) and multigrid tools, as proposed in
[5, 6].

It should be pointed out that the above two approaches (MFS for (5) and the
multi-elliptic interpolation (6)) do not differ essentially. The representation theorem
of the multi-elliptic interpolation (see [5,6]) guarantees that the solution of the multi-
elliptic interpolation (6) can be expressed in the form:

$$v(x) = w(x) + \sum_{j=1}^{N} \alpha_j \Phi(x - x_j),$$

where the function $w \in H_0^2(\Omega_0)$ satisfies the equation $(\Delta - c^2 I)(\Delta - \lambda^2 I)w = 0$
everywhere (including also the interpolation points) and Φ is the fundamental
solution of the operator $(\Delta - c^2 I)(\Delta - \lambda^2 I)$. Since the Bessel function K_0 is rapidly
decreasing, the regular part of the representation (i.e. w) can be omitted provided
that Ω_0 is large enough, and the representation becomes a simple MFS-like form.

In both variants of the approach (the direct multi-elliptic interpolation (6) or
the MFS applied to (5)), a crucial problem is the proper choice of the scaling
parameter c. If it is too small, the solution of (5) does not approximate u^*, the
solution of the original problem (1). If it is too large, numerical singularities appear
at the interpolation points. Numerical experience shows (see [8]) that the optimal
compromise is when c is inversely proportional to the characteristic distance of the
boundary interpolation points. In the next section, this question is investigated in
details.

4 Error Estimations

Throughout this section, *const* will always denote a constant with not necessarily
the same value in each occurrence, but independent of the scaling factor c as well
as the boundary interpolation points $x_1, \ldots, x_N$.

4.1 Preliminary Lemmas

First we prove estimations for the traces of some solutions of certain (multi-) elliptic
equations. For the sake of simplicity, we restrict ourselves to the case when the
domain Ω is a half-stripe: $\Omega = (0, L) \times (0, +\infty)$, $\Gamma = (0, L) \times \{0\}$, and the
functions defined on Ω are considered L-periodic with respect to the first variable.
This makes the proof of the next two lemmas simpler:

Lemma 1. *Suppose that* $(\Delta - c^2 I)v = 0$ *in* Ω. *Then*

$$\|\frac{\partial v}{\partial n}|_\Gamma\|^2_{H^{-1/2}(\Gamma)} \leq const \cdot \left(\|v|_\Gamma\|^2_{H^{1/2}(\Gamma)} + c^2 \cdot \|v|_\Gamma\|^2_{H^{-1/2}(\Gamma)}\right) \qquad (8)$$

and

$$\|\frac{\partial v}{\partial n}|_\Gamma\|^2_{H^{1/2}(\Gamma)} \leq const \cdot \left(\|v|_\Gamma\|^2_{H^{3/2}(\Gamma)} + c^2 \cdot \|v|_\Gamma\|^2_{H^{1/2}(\Gamma)}\right) \qquad (9)$$

Proof. Expressing $v|_\Gamma$ in terms of complex Fourier series: $v|_\Gamma = \sum_k \alpha_k e^{i\kappa x}$ (where $\kappa := \frac{2k\pi}{L}$), the function v has the form: $v(x, y) = \sum_k \alpha_k e^{-\sqrt{\kappa^2+c^2}y} e^{i\kappa x}$. Thus: $\frac{\partial v}{\partial n}|_\Gamma = \sum_k \alpha_k \sqrt{\kappa^2 + c^2}\, e^{i\kappa x}$, therefore:

$$\|\frac{\partial u}{\partial n}|_\Gamma\|^2_{H^{-1/2}(\Gamma)} \leq const \cdot \sum_k |\alpha_k|^2 \frac{\kappa^2 + c^2}{\sqrt{\kappa^2 + \lambda^2}} \leq$$

$$\leq const \cdot \sum_k |\alpha_k|^2 \frac{\kappa^2 + \lambda^2 + c^2 - \lambda^2}{\sqrt{\kappa^2 + \lambda^2}} \leq$$

$$\leq const \cdot \sum_k |\alpha_k|^2 \sqrt{\kappa^2 + \lambda^2} + const \cdot (c^2 - \lambda^2) \cdot \sum_k |\alpha_k|^2 \frac{1}{\sqrt{\kappa^2 + \lambda^2}},$$

which implies (8). Equation (9) can be proved in a similar way.

Lemma 2. *Suppose that* $(\Delta - \lambda^2 I)(\Delta - c^2 I)u = 0$ *in* Ω. *If* $\frac{\partial u}{\partial n}|_\Gamma = 0$, *then:*

$$\left|\int_\Gamma \frac{\partial \Delta u}{\partial n} \bar{u}\, d\Gamma\right| \leq const \cdot \left(\|u|_\Gamma\|^2_{H^{3/2}(\Gamma)} + c \cdot \|u|_\Gamma\|^2_{H^1(\Gamma)} + c^2 \cdot \|u|_\Gamma\|^2_{H^{1/2}(\Gamma)}\right)$$

$$(10)$$

On the other hand, if $u|_\Gamma = 0$, *then:*

$$\|\Delta u|_\Gamma\|^2_{H^{-1/2}(\Gamma)} \leq const \cdot \left(\|\frac{\partial u}{\partial n}|_\Gamma\|^2_{H^{1/2}(\Gamma)} + c^2 \cdot \|\frac{\partial u}{\partial n}|_\Gamma\|^2_{H^{-1/2}(\Gamma)}\right) \qquad (11)$$

Proof. First, suppose that $\frac{\partial u}{\partial n}|_\Gamma = 0$ and $u|_\Gamma = \sum_k \alpha_k e^{i\kappa x}$. Then

$$u(x, y) = \sum_k \alpha_k (A_k e^{-\sqrt{\kappa^2+\lambda^2}y} + B_k e^{-\sqrt{\kappa^2+c^2}y}) e^{i\kappa x},$$

where

$$A_k = \frac{\sqrt{\kappa^2 + c^2}}{\sqrt{\kappa^2 + c^2} - \sqrt{\kappa^2 + \lambda^2}}, \qquad B_k = -\frac{\sqrt{\kappa^2 + \lambda^2}}{\sqrt{\kappa^2 + c^2} - \sqrt{\kappa^2 + \lambda^2}}.$$

Therefore

$$\frac{\partial \Delta u}{\partial n}|_\Gamma = \sum_k \alpha_k \left(\lambda^2 A_k \sqrt{\kappa^2 + \lambda^2} + c^2 B_k \sqrt{\kappa^2 + c^2}\right) e^{i\kappa x}.$$

From Parseval's theorem:

$$\left| \int_\Gamma \frac{\partial \Delta u}{\partial n} \bar{u}\, d\Gamma \right| \leq const \cdot \sum_k |\alpha_k|^2 (\lambda^2 |A_k| \sqrt{\kappa^2 + \lambda^2} + c^2 |B_k| \sqrt{\kappa^2 + c^2})$$

Using elementary estimations, from here (10) follows.
Now suppose that $u|_\Gamma = 0$ and $\frac{\partial u}{\partial n}|_\Gamma = \sum_k \beta_k e^{i\kappa x}$. Then

$$u(x,y) = \sum_k \beta_k (A_k e^{-\sqrt{\kappa^2 + \lambda^2}\, y} + B_k e^{-\sqrt{\kappa^2 + c^2}\, y}) e^{i\kappa x},$$

where now

$$A_k = -\frac{1}{\sqrt{\kappa^2 + c^2} - \sqrt{\kappa^2 + \lambda^2}}, \qquad B_k = \frac{1}{\sqrt{\kappa^2 + c^2} - \sqrt{\kappa^2 + \lambda^2}}.$$

Therefore

$$\Delta u|_\Gamma = \sum_k \beta_k (\lambda^2 A_k + c^2 B_k) e^{i\kappa x}.$$

from which (11) follows.

From now on, we assume that $\Omega \subset \mathbb{R}^2$ is a bounded, smooth domain, and denote by $\|.\|_0$ the norm of $L_2(\Omega)$.

Lemma 3. *For arbitrary $u \in H^2(\Omega)$ and $c > 0$:*

$$\left\| \frac{\partial u}{\partial n}|_\Gamma \right\|_{H^{-1/2}(\Gamma)}^2 \leq const \cdot \frac{1}{c} \left(\|\Delta u\|_0^2 + c^2 \|\operatorname{grad} u\|_0^2 \right) \tag{12}$$

Proof. Green's theorem states that for any sufficiently regular v:

$$\int_\Gamma \frac{\partial u}{\partial n} v\, d\Gamma = \int_\Omega (\Delta u) v\, d\Omega + \int_\Omega \operatorname{grad} u \cdot \operatorname{grad} v\, d\Omega.$$

Suppose now that $(\Delta - c^2 I)v = 0$, where $v|_\Gamma \in H^{1/2}(\Gamma)$ is arbitrary. Then:

$$\int_\Gamma \frac{\partial u}{\partial n} v\, d\Gamma = \frac{1}{c^2} \cdot \left(\int_\Omega \Delta u \Delta v\, d\Omega + c^2 \int_\Omega \operatorname{grad} u \cdot \operatorname{grad} v\, d\Omega \right)$$

The expression in the brackets is a *scalar product* which generates a (semi)norm. From the Cauchy inequality:

$$\left| \int_\Gamma \frac{\partial u}{\partial n} v\, d\Gamma \right| = \frac{1}{c^2} \cdot \left(\|\Delta u\|_0^2 + c^2 \|\operatorname{grad} u\|_0^2 \right)^{1/2} \cdot \left(\|\Delta v\|_0^2 + c^2 \|\operatorname{grad} v\|_0^2 \right)^{1/2}$$

But $\|\Delta v\|_0^2 = c^4 \|v\|_0^2$, whence

$$\left| \int_\Gamma \frac{\partial u}{\partial n} v \, d\Gamma \right| = \frac{1}{c} \cdot \left(\|\Delta u\|_0^2 + c^2 \|\text{grad } u\|_0^2 \right)^{1/2} \cdot \left(c^2 \|v\|_0^2 + \|\text{grad } v\|_0^2 \right)^{1/2}$$

Using Green's formula again:

$$c^2 \|v\|_0^2 + \|\text{grad } v\|_0^2 = \int_\Omega v \Delta v \, d\Omega + \int_\Omega |\text{grad } v|^2 \, d\Omega = \int_\Gamma v \frac{\partial v}{\partial n} \, d\Gamma \leq$$

$$\leq \|v|_\Gamma\|_{H^{1/2}(\Gamma)} \cdot \|\frac{\partial v}{\partial n}|_\Gamma\|_{H^{-1/2}(\Gamma)} \leq const \cdot c \cdot \|v|_\Gamma\|_{H^{1/2}(\Gamma)}^2,$$

where we have used (8) and the trivial estimation $\|.\|_{H^{-1/2}(\Gamma)} \leq \|.\|_{H^{1/2}(\Gamma)}$. We have obtained that

$$\left| \int_\Gamma \frac{\partial u}{\partial n} v \, d\Gamma \right| = \frac{const}{\sqrt{c}} \cdot \left(\|\Delta u\|_0^2 + c^2 \|\text{grad } u\|_0^2 \right)^{1/2} \cdot \|v|_\Gamma\|_{H^{1/2}(\Gamma)},$$

which completes the proof.

4.2   Estimations for the Solutions of Multi-elliptic Equations

In this subsection, let $u \in H^2(\Omega)$ be a solution of the multi-elliptic equation

$$(\Delta - \lambda^2 I)(\Delta - c^2 I) u = 0 \tag{13}$$

Theorem 1. *Suppose that $u \in H^2(\Omega)$ is a solution of (13). If $u|_\Gamma - 0$ then*

$$\|u\|_0^2 \leq const \cdot \left(\frac{1}{c^4} \|\frac{\partial u}{\partial n}\|_{H^{1/2}(\Gamma)} + \frac{1}{c^2} \|\frac{\partial u}{\partial n}\|_{H^{-1/2}(\Gamma)} \right) \tag{14}$$

On the other hand, if $\frac{\partial u}{\partial n}|_\Gamma = 0$, then

$$\|u\|_0^2 \leq const \cdot \left(\frac{1}{c^4} \|u|_\Gamma\|_{H^{3/2}(\Gamma)} + \frac{1}{c^2} \|u|_\Gamma\|_{H^{1/2}(\Gamma)} + \|u|_\Gamma\|_{H^{-1/2}(\Gamma)} \right) \tag{15}$$

Proof. For arbitrary, sufficiently smooth functions u, v, for which $u|_\Gamma = 0, v|_\Gamma = 0$, Green's theorem implies that

$$\int_\Omega \left((\Delta - \lambda^2 I)\left(I - \frac{1}{c^2}\Delta \right) u \right) v \, d\Omega - \int_\Omega u \left((\Delta - \lambda^2 I)\left(I - \frac{1}{c^2}\Delta \right) v \right) d\Omega =$$

$$= \frac{1}{c^2} \int_\Gamma \Delta u \frac{\partial v}{\partial n} \, d\Gamma - \frac{1}{c^2} \int_\Gamma \frac{\partial u}{\partial n} \Delta v \, d\Gamma \tag{16}$$

Now let $u \in H^2(\Omega)$ be a solution of (13) and let $v \in H^2(\Omega)$ be the (unique) solution of

$$(\Delta - \lambda^2 I)\left(I - \frac{1}{c^2}\Delta\right)v = -u, \quad v|_\Gamma = 0, \quad (\Delta - \lambda^2 I)v|_\Gamma = 0 \qquad (17)$$

The boundary conditions taken on v imply that $\Delta v|_\Gamma = 0$. Thus, from (16), it follows that

$$\|u\|_0^2 = \frac{1}{c^2}\int_\Gamma \Delta u \frac{\partial v}{\partial n}\, d\Gamma \leq \frac{1}{c^2}\|\Delta u|_\Gamma\|_{H^{-1/2}(\Gamma)} \cdot \left\|\frac{\partial v}{\partial n}\right\|_{H^{1/2}(\Gamma)} \qquad (18)$$

Now we prove that

$$\left\|\frac{\partial v}{\partial n}\right\|_{H^{1/2}(\Gamma)} \leq const \cdot \|u\|_0, \qquad (19)$$

where the *const* is independent of c. Indeed, (17) can be split into a pair of Helmholtz problems

$$\Delta v - \lambda^2 v = w, \quad v|_\Gamma = 0$$

$$w - \frac{1}{c^2}\Delta w = -u, \quad w|_\Gamma = 0$$

Multiplying the second equation by w and integrating over Ω, we obtain:

$$\|w\|_0^2 + \frac{1}{c^2}\|\mathrm{grad}\, w\|_0^2 \leq \|u\|_0 \cdot \|w\|_0,$$

whence $\|w\|_0 \leq \|u\|_0$ independently of c. Now the first equation implies (19):

$$\left\|\frac{\partial v}{\partial n}\right\|_{H^{1/2}(\Gamma)} \leq const \cdot \|v\|_{H^2(\Omega)} \leq const \cdot \|w\|_0 \leq const \cdot \|u\|_0,$$

where we have used the trace theorem and the regularity theorem. Applying this estimation to (18), we have:

$$\|u\|_0^2 \leq const \cdot \frac{1}{c^2}\|\Delta u|_\Gamma\|_{H^{-1/2}(\Gamma)} \cdot \|u\|_0$$

(14) is now a consequence of (11).

To prove (15), let $v \in H^2(\Omega)$ be the (unique) solution of the Helmholtz problem

$$(\Delta - c^2 I)v = 0, \quad v|_\Gamma = u|_\Gamma.$$

Since v obviously satisfies (13), the difference $(u - v)$ also satisfies (13), and $(u - v)|_\Gamma = 0$. Furthermore:

$$\|u\|_0^2 \leq 2\|v\|_0^2 + 2\|u - v\|_0^2 \qquad (20)$$

The terms $\|v\|_0^2$ and $\|u-v\|_0^2$ can be estimated separately. Since $v = \frac{1}{c^2}\Delta v$:

$$\|v\|_0^2 = \frac{1}{c^2}\int_\Omega v\Delta v\, d\Omega = -\frac{1}{c^2}\int_\Omega |\operatorname{grad} v|^2\, d\Omega + \frac{1}{c^2}\int_\Gamma \frac{\partial v}{\partial n}v\, d\Gamma \le$$

$$\le \frac{1}{c^2}\|\frac{\partial v}{\partial n}\|_{H^{-1/2}(\Gamma)} \cdot \|v|_\Gamma\|_{H^{1/2}(\Gamma)} \le$$

$$\le \frac{const}{c^2}\left(\|v|_\Gamma\|_{H^{1/2}(\Gamma)}^2 + c^2\|v|_\Gamma\|_{H^{-1/2}(\Gamma)}^2\right)^{1/2}\cdot \|v|_\Gamma\|_{H^{1/2}(\Gamma)},$$

where we have applied (8). Since $v|_\Gamma = u|_\Gamma$, elementary estimations show that

$$\|v\|_0^2 \le \frac{const}{c^2}\left(\|u|_\Gamma\|_{H^{1/2}(\Gamma)}^2 + c\cdot\|u|_\Gamma\|_{H^{-1/2}(\Gamma)}\cdot\|u|_\Gamma\|_{H^{1/2}(\Gamma)}\right)$$

In the right-hand side, apply Young's inequality:

$$\|u|_\Gamma\|_{H^{-1/2}(\Gamma)}\cdot\|u|_\Gamma\|_{H^{1/2}(\Gamma)} \le \frac{c}{2}\cdot\|u|_\Gamma\|_{H^{-1/2}(\Gamma)}^2 + \frac{1}{2c}\cdot\|u|_\Gamma\|_{H^{1/2}(\Gamma)}^2,$$

which implies that

$$\|v\|_0^2 \le const\cdot\left(\|u|_\Gamma\|_{H^{-1/2}(\Gamma)}^2 + \frac{1}{c^2}\cdot\|u|_\Gamma\|_{H^{1/2}(\Gamma)}^2\right)$$

The term $\|u-v\|_0^2$ in (20) can be estimated by (14):

$$\|u-v\|_0^2 \le const\cdot\left(\frac{1}{c^4}\|\frac{\partial v}{\partial n}\|_{H^{1/2}(\Gamma)}^2 + \frac{1}{c^2}\|\frac{\partial v}{\partial n}\|_{H^{-1/2}(\Gamma)}^2\right),$$

since $\frac{\partial u}{\partial n}|_\Gamma = 0$. Now (15) is a consequence of (8) and (9), since $v|_\Gamma = u|_\Gamma$.

The theorem has two immediate corollaries:

Corollary 1. *If $u \in H^2(\Omega)$ is an arbitrary solution of (13), then*

$$\|u\|_0^2 \le const\cdot\left(\|u|_\Gamma\|_{H^{-1/2}(\Gamma)}^2 + \frac{1}{c^2}\|u|_\Gamma\|_{H^{1/2}(\Gamma)}^2 + \frac{1}{c^4}\|u|_\Gamma\|_{H^{3/2}(\Gamma)}^2 \right. \tag{21}$$

$$\left. + \frac{1}{c^2}\|\frac{\partial u}{\partial n}\|_{H^{-1/2}(\Gamma)}^2 + \frac{1}{c^4}\|\frac{\partial u}{\partial n}\|_{H^{1/2}(\Gamma)}^2\right)$$

Corollary 2. *If $u^* \in H^2(\Omega)$ is the (unique) solution of the original problem*

$$\Delta u^* - \lambda^2 u^* = 0 \quad \text{in } \Omega, \qquad u^*|_\Gamma = u_0 \in H^{3/2}(\Gamma),$$

and $u \in H^2(\Omega)$ is the (unique) solution of the bi-Helmholtz problem

$$(\Delta - c^2 I)(\Delta - \lambda^2 I)u = 0 \quad \text{in } \Omega, \quad u|_\Gamma = u_0, \quad \frac{\partial u}{\partial n}\Big|_\Gamma = 0 \qquad (22)$$

then

$$\|u - u^*\|_0^2 \leq \frac{const}{c^2} \cdot \|u_0\|_{H^{3/2}(\Gamma)}^2 \qquad (23)$$

Remark. The result of the above Corollary is similar to that of [9]. However, in contrast to [9], here a homogeneous problem is investigated with nonhomogeneous boundary condition.

The solution of the above bi-Helmholtz problem can also be estimated with respect to a stronger norm:

Theorem 2. *If $u \in H^2(\Omega)$ is the (unique) solution of (22), then*

$$\|\Delta u\|_0^2 + (\lambda^2 + c^2) \cdot \|\operatorname{grad} u\|_0^2 + \lambda^2 c^2 \|u\|_0^2 \leq const \cdot c^2 \cdot \|u_0\|_{H^{3/2}(\Gamma)}^2 \qquad (24)$$

Proof. Multiplying the Eq. (22) by u and integrating over Ω:

$$0 = \int_\Omega ((\Delta - \lambda^2 I)(\Delta - c^2 I)u)u \, d\Omega =$$

$$= \int_\Omega (\Delta\Delta u)u \, d\Omega - (\lambda^2 + c^2) \cdot \int_\Omega (\Delta u)u \, d\Omega + \lambda^2 c^2 \|u\|_0^2 =$$

$$= \|\Delta u\|_0^2 + \int_\Gamma \frac{\partial \Delta u}{\partial n} u \, d\Gamma + (\lambda^2 + c^2) \cdot \|\operatorname{grad} u\|_0^2 + \lambda^2 c^2 \|u\|_0^2$$

Now (24) is a consequence of (10) and the embedding theorems for the trace spaces.

4.3 Estimations for Boundary Functions

We need also estimations for functions which vanish in a set of boundary points.

Theorem 3. *If $u \in H^{3/2}(\Gamma)$ vanishes in the points $x_1, \ldots, x_N \in \Gamma$, then*

$$\|u\|_{H^0(\Gamma)}^2 \leq const \cdot h^3 \cdot \|u\|_{H^{3/2}(\Gamma)}^2 \qquad (25)$$

$$\|u\|_{H^{1/2}(\Gamma)}^2 \leq const \cdot h^2 \cdot \|u\|_{H^{3/2}(\Gamma)}^2 \qquad (26)$$

$$\|u\|_{H^1(\Gamma)}^2 \leq const \cdot h \cdot \|u\|_{H^{3/2}(\Gamma)}^2 \qquad (27)$$

where h is the maximal distance of the neighboring boundary points.

Proof. We can assume that $\Gamma = [0, L]$. For any $x \in [x_k, x_{k+1}]$:

$$u(x) = \int_{x_k}^{x} u'(t)\,dt,$$

whence, using the Cauchy-Schwarz inequality: $\|u\|^2_{H^0(\Gamma)} \leq const \cdot h^2 \cdot \|u'\|^2_{H^0(\Gamma)}$ which implies that:

$$\|u\|_{H^0(\Gamma)} \leq const \cdot h \cdot \|u\|_{H^1(\Gamma)} \tag{28}$$

Now we apply the classical interpolation theorem [2] which states that if $r \leq s \leq t$ ($r \neq t$) and $\theta = \frac{t-s}{t-r}$, then

$$\|u\|_{H^s(\Gamma)} \leq \|u\|^\theta_{H^r(\Gamma)} \cdot \|u\|^{1-\theta}_{H^t(\Gamma)}$$

In particular, set $r := 0$, $s := 1$, $t := 3/2$, then $\theta = 1/3$ and

$$\|u\|_{H^1(\Gamma)} \leq \|u\|^{1/3}_{H^0(\Gamma)} \cdot \|u\|^{2/3}_{H^{3/2}(\Gamma)}$$

From here and (28) and (25) follows. Equation (25) with the previous inequality implies (27). Finally, applying the interpolation theorem again with $r := 0$, $s := 1/2$, $t := 1$ (then $\theta = 1/2$), we obtain that

$$\|u\|_{H^{1/2}(\Gamma)} \leq \|u\|^{1/2}_{H^0(\Gamma)} \cdot \|u\|^{1/2}_{H^1(\Gamma)}$$

From here, (25) and (27), (26) also follows.

4.4   Error Estimations for the RMFS

Now we are able to prove the main result of the section. Consider again the original problem:

$$\Delta u^* - \lambda^2 u^* = 0 \text{ in } \Omega, \quad u^*|_\Gamma = u_0 \in H^{3/2}(\Gamma) \tag{29}$$

and the bi-Helmholtz type boundary interpolation problem:

$$(\Delta - c^2 I)(\Delta - \lambda^2 I)v = 0 \text{ in } \Omega_0 \setminus \{x_1, \ldots, x_N\} \tag{30}$$

$$v(x_k) = u_0(x_k) \quad (k = 1, \ldots, N)$$

$$v|_{\partial\Omega_0} = 0, \quad \frac{\partial v}{\partial n}\Big|_{\partial\Omega_0} = 0$$

where $x_1, \ldots, x_N \in \Gamma$ are the boundary interpolation points and h is the maximal distance of the neighboring interpolation points. $\Omega_0 \supset \bar{\Omega}$ is a larger domain in which the boundary interpolation is performed. Recall that if Ω_0 is large enough, v can be considered a MFS-type approximate solution; however, it can be constructed also by direct solution of (30) using quadtrees and multigrid techniques.

Theorem 4. *With the above notations:*

$$\|u^* - v\|_0^2 \le const \cdot \left(h^3 c^2 + h^2 + \frac{1}{c^2} + \frac{1}{c}\right) \cdot \|u_0\|_{H^{3/2}(\Gamma)}^2 \tag{31}$$

In particular, if the scaling constant c is defined to be inversely proportional to the distance h, then:

$$\|u^* - v\|_0^2 \le const \cdot h \cdot \|u_0\|_{H^{3/2}(\Gamma)}^2$$

Remark. The estimation is not sharp due to the applied rough estimations e.g. the imbedding theorems. Numerical experiences show better result [8]. See also the example in Sect. 5.

Proof. Let $u \in H_0^2(\Omega_0)$ be the (unique) solution of the bi-Helmholtz problem defined on the extended domain Ω_0:

$$(\Delta - c^2 I)(\Delta - \lambda^2 I)u = 0 \text{ in } \Omega_0 \setminus \Gamma, \quad u|_\Gamma = u_0, \quad \frac{\partial u}{\partial n}|_\Gamma = 0$$

$$u|_{\partial\Omega_0} = 0, \quad \frac{\partial u}{\partial n}|_{\partial\Omega_0} = 0$$

Obviously, $\|u^* - v\|_0 \le \|u^* - u\|_0 + \|u - v\|_0$. From (23):

$$\|u^* - u\|_0^2 \le \frac{const}{c^2} \cdot \|u_0\|_{H^{3/2}(\Gamma)}^2,$$

so that it is sufficient to estimate $\|u - v\|_0$. From (21) it follows that

$$\|u - v\|_0^2 \le const \cdot \left(\|(u - v)|_\Gamma\|_{H^{-1/2}(\Gamma)}^2 + \frac{1}{c^2}\|(u - v)|_\Gamma\|_{H^{1/2}(\Gamma)}^2 + \right.$$

$$\left. + \frac{1}{c^4}\|(u - v)|_\Gamma\|_{H^{3/2}(\Gamma)}^2 + \frac{1}{c^2}\|\frac{\partial(u - v)}{\partial n}\|_{H^{-1/2}(\Gamma)}^2 + \frac{1}{c^4}\|\frac{\partial(u - v)}{\partial n}\|_{H^{1/2}(\Gamma)}^2\right)$$

The terms containing $(u - v)|_\Gamma$ and $\frac{\partial(u-v)}{\partial n}$ can be estimated separately. Since $(u - v)|_\Gamma$ vanishes in all boundary interpolation points, Theorem 3 can be applied and we obtain:

$$\|(u - v)|_\Gamma\|_{H^{-1/2}(\Gamma)}^2 \le const \cdot h^3 \cdot \|(u - v)|_\Gamma\|_{H^{3/2}(\Gamma)}^2$$

and

$$\|(u - v)|_\Gamma\|_{H^{1/2}(\Gamma)}^2 \le const \cdot h^2 \cdot \|(u - v)|_\Gamma\|_{H^{3/2}(\Gamma)}^2$$

Thus, the trace theorem implies that:

$$\|(u-v)|_\Gamma\|^2_{H^{-1/2}(\Gamma)} + \frac{1}{c^2}\|(u-v)|_\Gamma\|^2_{H^{1/2}(\Gamma)} + \frac{1}{c^4}\|(u-v)|_\Gamma\|^2_{H^{3/2}(\Gamma)} \le$$

$$\le const \cdot \left(h^3 + \frac{h^2}{c^2} + \frac{1}{c^4}\right) \cdot \|u-v\|^2_{H^2(\Omega)} \le$$

$$\le const \cdot \left(h^3 + \frac{h^2}{c^2} + \frac{1}{c^4}\right) \cdot \|u-v\|^2_{H^2(\Omega_0)}$$

In the subspace $H_0^2(\Omega_0)$, the $H^2(\Omega_0)$-norm is equivalent to the norm $\|w\| := \|\Delta w\|_{L_2(\Omega_0)}$, whence

$$\|u-v\|^2_{H^2(\Omega_0)} \le const \cdot \left(\|\Delta(u-v)\|^2_{L_2(\Omega_0)} + \right.$$

$$\left. + (\lambda^2 + c^2) \cdot \|grad\,(u-v)\|^2_{L_2(\Omega_0)} + \lambda^2 c^2 \|u-v\|^2_{L_2(\Omega_0)}\right) \le$$

$$\le const \cdot \left(\|\Delta u\|^2_{L_2(\Omega_0)} + (\lambda^2 + c^2) \cdot \|grad\,u\|^2_{L_2(\Omega_0)} + \lambda^2 c^2 \|u\|^2_{L_2(\Omega_0)}\right)$$

where we utilized the fact that $(u-v)$ is the orthogonal projection of u (see Sect. 3). From here and (24), it follows that

$$\|u-v\|^2_{H^2(\Omega_0)} \le const \cdot c^2 \cdot \|u_0\|^2_{H^{3/2}(\Gamma)},$$

that is:

$$\|(u-v)|_\Gamma\|^2_{H^{-1/2}(\Gamma)} + \frac{1}{c^2}\|(u-v)|_\Gamma\|^2_{H^{1/2}(\Gamma)} + \frac{1}{c^4}\|(u-v)|_\Gamma\|^2_{H^{3/2}(\Gamma)} \le$$

$$\le const \cdot \left(h^3 c^2 + h^2 + \frac{1}{c^2}\right) \cdot \|u_0\|^2_{H^{3/2}(\Gamma)}.$$

The term $\|\frac{\partial(u-v)}{\partial n}\|^2_{H^{1/2}(\Gamma)}$ can be estimated by the trace theorem:

$$\|\frac{\partial(u-v)}{\partial n}\|^2_{H^{1/2}(\Gamma)} \le const \cdot \|u-v\|^2_{H^2(\Omega_0)},$$

while the term $\|\frac{\partial(u-v)}{\partial n}\|^2_{H^{-1/2}(\Gamma)}$ can be estimated by (12):

$$\|\frac{\partial(u-v)}{\partial n}\|^2_{H^{-1/2}(\Gamma)} \le \frac{const}{c} \cdot \left(\|\Delta(u-v)\|^2_{L_2(\Omega_0)} + c^2\|grad\,(u-v)\|^2_{L_2(\Omega_0)}\right).$$

Consequently:

$$\frac{1}{c^2}\left\|\frac{\partial(u-v)}{\partial n}\right\|^2_{H^{-1/2}(\Gamma)} + \frac{1}{c^4}\left\|\frac{\partial(u-v)}{\partial n}\right\|^2_{H^{1/2}(\Gamma)} \leq$$

$$\leq const \cdot \left(\frac{1}{c^3} \cdot \left(\|\Delta(u-v)\|^2_{L_2(\Omega_0)} + c^2\|\text{grad}\,(u-v)\|^2_{L_2(\Omega_0)}\right) + \right.$$

$$\left. + \frac{1}{c^4} \cdot \|u-v\|^2_{H^2(\Omega_0)}\right) \leq$$

$$\leq const \cdot \left(\frac{1}{c^3} + \frac{1}{c^4}\right) \cdot \left(\|\Delta u\|^2_{L_2(\Omega_0)} + (\lambda^2 + c^2) \cdot \|\text{grad}\,u\|^2_{L_2(\Omega_0)} + \right.$$

$$\left. + \lambda^2 c^2 \|u\|^2_{L_2(\Omega_0)}\right) \leq$$

$$\leq const \cdot \left(\frac{1}{c} + \frac{1}{c^2}\right) \cdot \|u_0\|^2_{H^{3/2}(\Gamma)},$$

where we applied the same variational arguments as above. This completes the proof.

5   A Numerical Example

Consider again the problem (1) with the exact test solution (4):

$$u^*(x, y) = (\sin x) \cdot \sinh(\sqrt{1 + \lambda^2} \cdot y),$$

where $\lambda := 2$, and the domain of the problem is the unit circle again. The equation is supplied with Dirichlet boundary condition in a consistent way. Table 3 shows the relative L_2-errors of the approximate solutions obtained by the regularized MFS-technique, using the fundamental solution of the bi-Helmholtz operator $(\Delta - c^2 I)(\Delta - \lambda^2 I)$ with different numbers of equally spaced boundary collocation points (N) and scaling factors (c). It can clearly be seen that the optimal scaling factor is proportional to the number of boundary collocation points i.e. inversely proportional to the characteristic distance of the boundary points. It should also be pointed out that, in contrast to the traditional version of the Method of Fundamental Solutions, the condition number of the linear system (3) remains moderate due to the fact that now the source points coincide with the boundary collocation points. Table 4 shows the condition numbers in the case of the optimal choice of the scaling parameter c. Comparing the results with that of Tables 1 and 2, the RMFS produces significantly less exact approximations, but the linear systems to be solved are much better conditioned.

Table 3 Relative L_2-errors (%) of the RMFS-solution of the Helmholtz test problem. N is the number of boundary collocation points, c is the scaling factor

$N \setminus c$	8	16	32	64	128	256	512
16	18.45	**8.47**	20.61	32.93	42.32	49.51	55.14
32	24.30	12.49	**3.01**	10.34	19.10	26.49	32.68
64	25.24	14.99	7.10	**1.37**	5.30	10.55	15.30
128	25.36	15.36	8.22	3.75	**0.65**	2.70	5.57
256	25.37	15.41	8.39	4.29	1.93	**0.27**	1.30

Table 4 Condition numbers of the RMFS-matrices, modified Helmholtz test problem. N is the number of boundary collocation points, the scaling factor c is optimal

N	16	32	64	128	256	512
Condition number	3.332	6.519	13.01	26.05	52.16	104.4

6 Summary and Conclusions

A regularized version of the Method of Fundamental Solutions was presented. In contrast to the traditional approach, the original elliptic problem is approximated first a fourth-order multi-elliptic one controlled by a scaling parameter. This problem can be treated by the Method of Fundamental Solutions without difficulty. In addition to it, since the fundamental solution of the new problem is continuous at the origin, the source points appearing in the Method of Fundamental Solutions can be located on the boundary without generating numerical singularities in the approximate solution. The use of the MFS can be avoided by performing a multi-elliptic boundary interpolation, which circumvents the problem of the use of dense and ill-conditioned matrices. Applying the tools of the multi-elliptic interpolation, a priori error estimations have been proved. The optimal choice of the scaling factor appearing in the multi-elliptic approximation has turned out to be inversely proportional to the characteristic distance of the boundary interpolation points. In the presented method, the source and the collocation point are allowed to coincide, without generating singularities, so that no desingularization technique is needed. The exactness of the RMFS is significantly lower than that of the traditional MFS, however, the appearing linear systems are much better conditioned. As a serious limitation, the technique in this form can be applied to the case of Dirichlet boundary condition only.

Acknowledgements The research was partly supported by the European Union (co-financed by the European Social Fund) under the project TÁMOP 4.2.1/B-09/1/KMR-2010-0003.

References

1. C.J.S. Alves, C.S. Chen, B. Sarler, *The Method of Fundamental Solutions for Solving Poisson Problems*, ed. by C.A. Brebbia, A. Tadeu, V. Popov. International Series on Advances in Boundary Elements, vol. 13 (WIT, Southampton, 2002), pp. 67–76
2. I. Babuška, A.K. Aziz, *Survey Lectures on the Mathematical Foundations of the Finite Element Method*, ed. by A.K. Aziz. The Mathematical Foundations of the Finite Element Method with Applications to Partial Differential Equations (Academic, New York, 1972)
3. W. Chen, L.J. Shen, Z.J. Shen, G.W. Yuan, Boundary knot method for poisson equations. Eng. Anal. Boundary Elem. **29/8**, 756–760 (2005)
4. W. Chen, F.Z. Wang, A method of fundamental solutions without fictitious boundary. Eng. Anal. Boundary Elem. **34**, 530–532 (2010)
5. C. Gáspár, Multi-level biharmonic and bi-Helmholtz interpolation with application to the boundary element method. Eng. Anal. Boundary Elem. **24/7–8**, 559–573 (2000)
6. C. Gáspár, *Fast Multi-level Meshless Methods Based on the Implicit Use of Radial Basis Functions*, ed. by M. Griebel, M.A. Schweitzer. Lecture Notes in Computational Science and Engineering, Vol. 26 (Springer, New York, 2002), pp. 143–160.
7. C. Gáspár, *A Multi-Level Regularized Version of the Method of Fundamental Solutions*, ed. by C.S. Chen, A. Karageorghis, Y.S. Smyrlis. The Method of Fundamental Solutions—A Meshless Method (Dynamic, Atlanta, 2008), pp. 145–164
8. C. Gáspár, Multi-level meshless methods based on direct multi-elliptic interpolation. J. Comput. Appl. Math. **226/2**, 259–267 (2009). Special issue on large scale scientific computations
9. T.K. Nilssen, X.-C. Tai, R. Winther, A robust nonconforming H^2-Element. Math. Comput. **70/234**, 489–505 (2000)
10. B. Šarler, Solution of potential flow problems by the modified method of fundamental solutions: formulations with the single layer and the double layer fundamental solutions. Eng. Anal. Boundary Elem. **33**, 1374–1382 (2009)
11. D.L. Young, K.H. Chen, C.W. Lee, Novel meshless method for solving the potential problems with arbitrary domain. J. Computat. Phys. **209**, 290–321 (2005)

A Characteristic Particle Method for Traffic Flow Simulations on Highway Networks

Yossi Farjoun and Benjamin Seibold

Abstract A characteristic particle method for the simulation of first order macroscopic traffic models on road networks is presented. The approach is based on the method *particleclaw*, which solves scalar one dimensional hyperbolic conservation laws exactly, except for a small error right around shocks. The method is generalized to nonlinear network flows, where particle approximations on the edges are suitably coupled together at the network nodes. It is demonstrated in numerical examples that the resulting particle method can approximate traffic jams accurately, while only devoting a few degrees of freedom to each edge of the network.

Keywords Particle • Characteristic • Particleclaw • Shock • Traffic flow • Network

1 Introduction

Macroscopic traffic models describe the evolution of the vehicular traffic density on a road by partial differential equations. A large network of highways, which includes ramps and intersections, can be modeled as a directed graph, whose edges are road segments that join and bifurcate at network nodes. The traffic density on each edge evolves according to some macroscopic traffic model, and at nodes,

Y. Farjoun (✉)
Broad Institute of MIT and Harvard, 7 Cambridge Center, Cambridge, MA 02142
e-mail: farjoun@broadinstitute.org

B. Seibold
Department of Mathematics, Temple University, 1805 North Broad Street, Philadelphia, PA 19122, USA
e-mail: seibold@temple.edu

M. Griebel and M.A. Schweitzer (eds.), *Meshfree Methods for Partial Differential Equations VI*, Lecture Notes in Computational Science and Engineering 89, DOI 10.1007/978-3-642-32979-1_13, © Springer-Verlag Berlin Heidelberg 2013

specific coupling conditions are specified. Since realistic road networks can easily consist of tens on thousands of edges, efficient numerical methods are crucial for the simulation, forecasting, now-casting, or optimization of traffic flow on networks.

In this paper, we present a characteristic particle method for "first order" traffic models on road networks. The approach is based on the method *particleclaw* [11, 30], which solves scalar one dimensional hyperbolic conservation laws exactly, except for the immediate vicinity of shocks, where a small approximation error is observed. Particleclaw is used to evolve the numerical solution on each edge, and a methodology is presented for the coupling of the particle approximations on individual edges together through the network nodes. In recent years [10, 12, 13], particleclaw has been demonstrated to possess certain structural advantages over approaches that are based on a fixed grid. To name a few examples:

- It possesses no numerical viscosity, and thus preserves a reasonable accuracy even with very few degrees of freedom. This is in contrast to Godunov's method [15] and other low order schemes.
- It is optimally local, in the sense that particles move independently of each other, and communication between neighboring particles occurs only at shocks. In contrast, high order finite difference approaches [24], finite volume methods [31], or ENO [17]/WENO [26] schemes use wide stencils to achieve high order, which poses a particular challenge right at network nodes.
- It is total variation diminishing, yet it is second order accurate even in the presence of shocks [11]. In contrast, many limiters in fixed grid approaches need to reduce the order of convergence near shocks in order to avoid overshoots.
- The approximate solution is represented as a piecewise similarity solution of the underlying equation, that is continuous except for right at actual shocks. In contrast, the reconstruction in finite volume [31] and Discontinuous Galerkin [6, 28] methods tends to possess spurious discontinuities at every boundary between cells.
- It is adaptive by construction. Particles adapt to the shape of the solution, and the approach can be generalized to incorporate stiff source terms [13]. This is in contrast to fixed grid methods, which would need to use mesh refinement techniques [3] for adaptivity.

These advantages render *particleclaw* an interesting candidate for the simulation of traffic flow on networks. The need for the design of specialized numerical schemes for traffic models on networks has been pointed out for instance by Bretti et al. [4], who design an efficient numerical approach for a very specific class of flux functions, for which it suffices to track the location of the transition between free and congested traffic flow. The idea of tracking features is also shared by front tracking methods [21], which approximate the whole solution by finitely many moving discontinuities. In fact, in analogy to front tracking, *particleclaw* can also be interpreted as a *rarefaction tracking* [12].

This paper is organized as follows. In Sect. 2, the class of traffic models under consideration is outlined, and the coupling conditions for road networks are given. The characteristic particle method *particleclaw* is presented in Sect. 3, and in

Sect. 4, we demonstrate how the approach can be generalized to nonlinear flows on networks. Numerical results are shown in Sect. 5, and in Sect. 6, we present conclusions and give an outlook.

2   Traffic Models on Highway Networks

2.1   Macroscopic Traffic Models

Macroscopic traffic models treat the traffic as a continuum, and use partial differential equations to describe the temporal evolution of the vehicle density $u(x,t)$,[1] where x is the position on the road (the flow is averaged over all lanes that go in one direction), and t is time. If vehicles move with the velocity $v(x,t)$, then the conservation of vehicles (in the absence of ramps) is described by the continuity equation $u_t + (uv)_x = 0$. The assumption of a direct functional relationship between the velocity and the density, $v = v(u)$, yields the classical Lighthill-Whitham-Richards (LWR) model [25, 29]

$$u_t + (f(u))_x = 0 \, , \tag{1}$$

which is a scalar hyperbolic conservation law with a flux function $f = uv$ that equals the traffic flow rate. Due to the direct density-velocity relationship, the LWR model does not model vehicle accelerations, and thus is not able to describe non-equilibrium features such as phantom traffic jams or self-sustained traffic waves ("jamitons") [14]. To overcome this limitation, numerous types of "second order" models have been developed, such as the Payne-Whitham model [27], the Zhang-Aw-Rascle model [2, 32], phase transition models [8], and many more. These introduce a velocity (or velocity-like) variable as an independent unknown into the equations, resulting in a system of balance laws. In this paper, we restrict our study to effects that can be captured reasonably well by the scalar LWR model, i.e. we are interested in the large scale, nonlinear equilibrium behavior of traffic flow.

As it is common in traffic models [18], we assume that the LWR flux function satisfies the following conditions: (i) no flow for empty road and bumper-to-bumper densities, i.e. $f(0) = 0 = f(u^m)$, for some maximum density u^m; and (ii) concavity, i.e. $f''(u) < 0 \ \forall u \in [0, u^m]$. As a consequence of these assumptions, there is a unique maximum flow rate $f^* = f(u^*)$ that occurs for an optimal density u^*. Two parameters of f are fundamentally important: the slope $v^m = f'(0)$ is the velocity of vehicles when alone on the road (i.e. approximately the speed limit);

---

[1]Even though densities are commonly denoted by ρ, here we use u, in order to express the fact that the numerical approach applies to more general network flows.

and u^m is the number of lanes divided by the average vehicle length plus a safety distance. Other than for these two values, the precise functional shape of f can depend on many factors, such as the road geometry, the drivers' state of mind, etc. Frequently, one assumes a simple parabolic profile $f(u) = v^m u (1 - \frac{u}{u^m})$. This particular shape was inspired by measurements of Greenshields [16], and even though it does not fit well with contemporary measurements [18], it is commonly used due to its simplicity. We shall do so here as well, as it simplifies the presentation of the numerical approach in Sect. 4.

2.2 Traffic Networks

A traffic network is a directed graph of *network edges* (road segments) that join and bifurcate at *network nodes*. On each edge, the traffic evolves according to the LWR model (1) with a flux function that is specific to this edge (here, we assume that on each edge the flux function does not explicitly depend on space or time). In order to have a feasible model for traffic flow on road networks, one must formulate coupling conditions that guarantee the existence of a unique entropy solution, given suitable initial and boundary conditions. This problem was first addressed by Holden and Risebro [22], and subsequently generalized by various other authors (see below). In the following, we briefly outline the key ideas of coupling conditions at network nodes.

At a network node, one has the following setup. Let the node have n ingoing edges (roads), numbered $i = 1, 2, \ldots, n$, each of which carries a vehicle density $u_i(x_i, t)$, where $x_i \in [0, L_i]$, and a flux function $f_i(u)$. Similarly, let the node have m outgoing edges, numbered $j = n+1, \ldots, n+m$, each carrying a density $u_j(x_j, t)$, where $x_j \in [0, L_j]$, and a flux function $f_j(u)$. Ingoing edges end at the node at $x_i = L_i$, and outgoing edges start at the node at $x_j = 0$. The conservation of vehicles requires that the total inflow into the node equals the total outflow out of the node, i.e.

$$\sum_{i=1}^{n} f_i(u_i(L_i, t)) = \sum_{j=n+1}^{n+m} f_j(u_j(0, t)) \ \forall t \ . \tag{2}$$

In order to obtain further rules for the temporal evolution of the solution at nodes, a generalized Riemann problem is formulated, as follows. Let all edges be extended away from the node to $\pm\infty$, and consider initial conditions that are constant on each edge, i.e. $u_i(x, 0) = u_i$, for $x \in [-\infty, L_i]$, and $u_j(x, 0) = u_j$ for $x \in [0, \infty]$. The question is then: what are the new states $\hat{u}_i = u_i(L_i, t)$ and $\hat{u}_j = u_j(0, t)$ at the node for $t > 0$? These new states define the solution at all times, since the problem admits a self-similar solution $u_i(x, t) = w_i(\frac{L_i - x}{t})$ and $u_j(x, t) = w_j(\frac{x}{t})$. Let the flux values of the new states be denoted $\gamma_i = f_i(\hat{u}_i) \ \forall \ i \in \{1, \ldots, n+m\}$. Clearly, these fluxes must satisfy the conservation condition (2), i.e.

$$\sum_{i=1}^{n} \gamma_i = \sum_{j=n+1}^{n+m} \gamma_j . \tag{3}$$

Moreover, the new states $\hat{u}_i \; \forall \, i \, \in \, \{1, \ldots, n + m\}$ must generate a solution for which information on each edge is either stationary or moves away from the node (information cannot move "into" the node). As shown in [22], this information direction requirement implies that the new fluxes satisfy the inequality constraints

$$\gamma_i \in \Omega_i \text{ , where } \Omega_i = [0, f_i(\min\{u_i, u_i^*\})] \; \forall \, i \in \{1, \ldots, n\} \text{ , and} \tag{4}$$

$$\gamma_j \in \Omega_j \text{ , where } \Omega_j = [0, f_j(\max\{u_j, u_j^*\})] \; \forall \, j \in \{n + 1, \ldots, n + m\} . \tag{5}$$

In words: on ingoing edges, the new flux must be less than or equal to the *demand flux*, which is given by the old flux if $u_i \le u_i^*$, and the maximum flux f_i^* if $u_i > u_i^*$. Likewise, on outgoing edges, the new flux must be less than or equal to the *supply flux*, which is given by the old flux if $u_j \ge u_j^*$, and the maximum flux f_j^* if $u_j < u_j^*$.

Conditions (3), (4), and (5) allow infinitely many possible new fluxes. Further conditions must be provided to single out a unique solution. A reasonable set of additional conditions is given by the drivers' desired destinations matrix, defined as follows. Let γ_{in} denote the vector of ingoing fluxes, and γ_{out} be the vector of outgoing fluxes. The fact that very specific percentages of drivers that enter the node through edge i exit the node on edge j (different drivers have different destinations) can be encoded via the linear system

$$A \cdot \gamma_{\text{in}} = \gamma_{\text{out}} , \tag{6}$$

where

$$A = \begin{pmatrix} a_{1,n+1} & \cdots & a_{n,n+1} \\ \vdots & \ddots & \vdots \\ a_{1,n+m} & \cdots & a_{n,n+m} \end{pmatrix}, \quad \gamma_{\text{in}} = \begin{pmatrix} \gamma_1 \\ \vdots \\ \gamma_n \end{pmatrix}, \quad \gamma_{\text{out}} = \begin{pmatrix} \gamma_{n+1} \\ \vdots \\ \gamma_{n+m} \end{pmatrix}.$$

The matrix A is column-stochastic, i.e. all $a_{i,j} \in [0, 1]$ and $\sum_{j=n+1}^{n+m} a_{i,j} = 1$. Thus, the flux balance condition (3) is automatically guaranteed by relation (6), since $e^T \cdot \gamma_{\text{out}} = e^T \cdot A \cdot \gamma_{\text{in}} = e^T \cdot \gamma_{\text{in}}$, where $e = (1, \ldots, 1)^T$. Condition (6) together with the constraints (4) and (5) yields that γ_{in} must lie in the *feasibility domain*

$$\Omega = \{\gamma \in \Omega_1 \times \cdots \times \Omega_n \mid A \cdot \gamma \in \Omega_{n+1} \times \cdots \times \Omega_{n+m}\} \subset \mathbb{R}^n . \tag{7}$$

The selection of a specific $\gamma_{\text{in}} \in \Omega$ requires further modeling arguments. Possible criteria are, for instance, entropy arguments [22], the modeling of the intersection geometry [19], or simply the assumption that drivers behave such that the throughput through the node is maximized [5,7], i.e. one solves the linear program

$$\max e^T \cdot \boldsymbol{\gamma}_{\text{in}} \text{ s.t. } \boldsymbol{\gamma}_{\text{in}} \in \Omega . \tag{8}$$

In the examples presented in Sect. 5, we shall follow the latter option, even though
the other alternatives are possible as well. The modeling has to be augmented by one
small but important detail. It is possible that (8) does not possess a unique solution,
namely if the extremal boundary of Ω is perpendicular to e. In this case, one can
introduce the additional constraint that $\boldsymbol{\gamma}_{\text{in}} \in c\mathbb{R}$, where $c \in \mathbb{R}^n$ is a given constant
that models the merging behavior at the node.

The definition of the generalized Riemann problem is finalized by selecting new
states as follows. On ingoing roads, choose $\hat{u}_i = u_i$ if $\gamma_i = f_i(u_i)$, otherwise
choose $\hat{u}_i \geq u_i^*$, s.t. $f_i(\hat{u}_i) = \gamma_i$. Similarly, on outgoing roads, choose $\hat{u}_j = u_j$ if
$\gamma_j = f_j(u_j)$, otherwise choose $\hat{u}_j \leq u_j^*$, s.t. $f_j(\hat{u}_j) = \gamma_j$. By construction of (4)
and (5), any resulting shocks and rarefaction waves are guaranteed to move away
from the node (i.e. forward on outgoing and backward in ingoing edges).

3 Particleclaw

3.1 Characteristic Particles and Similarity Interpolant

On each network edge, we have a scalar one-dimensional hyperbolic conservation
law

$$u_t + (f(u))_x = 0 , \quad u(x,0) = u_0(x) , \tag{9}$$

where the flux function f is assumed to be twice differentiable and concave
($f'' < 0$) on the range of function values (see [11] for extensions of the approach
to flux functions with inflection points). We consider a special subset of exact
solutions of (9), which can be represented by a finite number of characteristic
particles, as follows. A particle is a computational node that carries a position $x^i(t)$,
and a function value $u^i(t)$. Note that we shall consistently denote particle indices
by superscripts, while subscripts are reserved for edge indices. In the following,
for convenience, we omit the time-dependence in the notation. Consider a time-
dependent set of n particles $P = \{(x^1, u^1), \dots, (x^n, u^n)\}$, where $x^1 \leq \dots \leq x^n$.
On the interval $[x^1, x^n]$ that is spanned by the particles, we define the *similarity
interpolant* $U_P(x)$ piecewise between neighboring particles as a true similarity
solution of (9). If $u^i \neq u^{i+1}$, it is implicitly given (and uniquely defined, since
f is concave) by

$$\frac{x - x^i}{x^{i+1} - x^i} = \frac{f'(U_P(x)) - f'(u^i)}{f'(u^{i+1}) - f'(u^i)} . \tag{10}$$

If $u^i = u^{i+1}$, the interpolant is simply constant $U_P(x) = u^i$. As shown in [11], the
interpolant U_P is an analytical solution of the conservation law (9), as each particle
moves according to the characteristic equations

$$\dot{x} = f'(u), \quad \dot{u} = 0, \tag{11}$$

i.e. $(x^i(t), u^i(t)) = (x^i(0) + f'(u^i(0))t, u^i(0))$. The reason for this fact is that due to the particular form of U_P, each point $(x(t), u(t))$ on it does move according to the same characteristic equations (11). Note that strictly speaking, $U_P(x, t)$ is a solution only in the weak sense, since the derivative $\frac{\partial}{\partial x} U_P(x, t)$ is discontinuous at the particles. However, since U_P is continuous, the Rankine-Hugoniot shock conditions [9] are trivially satisfied.

The interpolant U_P is called "similarity interpolant", since the solution between neighboring particles (10) is either a rarefaction wave that comes from a discontinuity (if $f'(x^i) < f'(x^{i+1})$) or a compression wave that will become a shock (if $f'(x^i) > f'(x^{i+1})$). As a consequence, the approach can be interpreted as *rarefaction tracking* [12]. This expresses both its similarities and differences to front tracking approaches [21,23], which approximate the true solution by a finite number of shocks.

Just as the true solution of (9) may cease to be continuous after some critical time, the similarity approximation exists only up to the time of the first collision of neighboring particles. For a pair of neighboring particles (x^i, u^i) and (x^{i+1}, u^{i+1}), the time of collision (i.e. they have the same x-position) is given by

$$T^i = -\frac{x^{i+1} - x^i}{f'(u^{i+1}) - f'(u^i)}, \tag{12}$$

given that $f'(x^i) > f'(x^{i+1})$. Consequently, for n particles, the time of the first collision is $T^* = \min(\{T^i \mid T^i > 0\} \cup \infty)$. At that time, a shock occurs (at $x^i = x^{i+1}$, from u^{i+1} to u^i), and the method of characteristics cannot be used to evolve the solution further in time.

3.2 Representation of Shocks

The characteristic motion of particles, as described in Sect. 3.1, can only be performed if no shocks are present in the numerical solution. One idea to overcome this limitation, thus admitting an evolution of solutions with shocks, is to merge particles upon collision. This approach was first presented in [10], analyzed and generalized in [11, 12], and implemented in the software *particleclaw* [30]. The merging of two particles (x^i, u^i) and (x^{i+1}, u^{i+1}) with $x^i = x^{i+1}$ into a single new particle $(x^i, \bar{u}^i)$ is performed such that the total area under the similarity interpolant U_P is exactly preserved. As shown in [11], the area under U_P between two neighboring particles equals

$$\int_{x^i}^{x^{i+1}} U_P(x) \, dx = (x^{i+1} - x^i) a(u^i, u^{i+1}), \tag{13}$$

where

$$a(u^i, u^{i+1}) = \frac{[f'(u)u - f(u)]_{u^i}^{u^{i+1}}}{[f'(u)]_{u^i}^{u^{i+1}}} \qquad (14)$$

is a nonlinear average function (note that $[f(u)]_v^w = f(w) - f(v)$). Equating the area under the interpolant before and after the merge, in general, yields a nonlinear equation for $\bar{u}^i$, which can be solved up to machine accuracy by a few Newton iteration steps (the geometry of the problem generally yields a very good initial guess). In the case we are solving here, the interpolant is linear which simplifies many calculations.

The merging of colliding particles replaces a discontinuity by a continuous interpolant, and thus the numerical approximation can be evolved further in time using the characteristic particle motion (11). Since the merging of particles i and $i + 1$ modifies the interpolant in the interval $[x^{i-1}, x^{i+2}]$, this approach introduces a small error right around shocks. We control the magnitude of this error by the following additional step. Let a parameter d be given on the edge that provides an upper bound on the distance of particles adjacent to a pair of particles that need to be merged. Consider two particles i and $i + 1$ that need to be merged, because $x^i = x^{i+1}$ and $f'(u^i) > f'(u^{i+1})$. If $x^i - x^{i-1} > d$, then a new particle is inserted (on the interpolant) at $x^i - d$. Moreover, if $x^{x+2} - x^{x+1} > d$, then a new particle is inserted at $x^{i+1} + d$. After the relabeling of the particle indices, the original particles i and $i + 1$ are merged, in the way described above.

Note that there is an alternative to particle merges: the generalization to shock particles, as introduced in [13]. The advantage of this version of the particle method is that shocks are represented exactly. The price to pay is a particle evolution that is more complicated than (11). In this paper, for the application to macroscopic traffic models on networks, we consider the approach that uses characteristic particles. While it is certainly possible to formulate the ideas presented below with shock particles, the simplicity of the characteristic particle motion admits an easier presentation and analysis of the methodology. Moreover, the approach presented here introduces an intrinsic error at network vertices, and thus the exact nature of shock particles is less of an advantage.

4 Generalizing Particleclaw to Network Flows

In this section we demonstrate how the approach *particleclaw* can be generalized to nonlinear traffic models on highway networks. Our goal is to obtain an overall approach that is highly modular, in the sense that the solution on each edge can be evolved independently of the other edges, and edges communicate only during a *synchronization step*. The specific methodology is as follows. Having sampled the initial conditions on each edge and synchronized them (as described below), we pick a time step Δt. During each time increment $t \in [t_n, t_n + \Delta t]$, the solution on each edge is evolved using the simple particle method described in Sect. 3, with a special treatment at the first and the last particle (see Sect. 4.1). At the end of the time step, at each network node, the adjacent edges are synchronized with each other: first, the numerical solution (which may have partially moved away from the edge) is interpolated/extrapolated back onto the edge; second, the generalized Riemann problems (described in Sect. 2.2) are invoked; lastly, area is suitably re-distributed. All these operations are performed such that the approach is exactly conservative, i.e. no area is lost or created.

The independent evolution of the solution on the edges renders the approach extremely adapt towards parallelization: each edge can be stored in its own share of memory, communication between the different edges need only occur during the synchronization, and very little information must be transferred (see Sect. 4.2). This methodology is possible due to the finite speed of information propagation of the hyperbolic conservation law (1) on each edge. There is a maximum synchronization time step Δt_{max}, such that for all $\Delta t \leq \Delta t_{max}$, information does not propagate further than half the length of each edge. As a consequence, in the synchronization step, all nodes can be treated independently of each other. Note that the maximum admissible time step Δt_{max} between synchronization events is on the order of the smallest edge length divided by the fastest characteristic velocity. This is significantly larger than the maximum admissible time step in many traditional numerical approaches, which is on the order of the grid resolution divided by the fastest characteristic velocity. Hence with the presented particle approach, the relatively costly generalized Riemann problems need to be called much less frequently.

It should be pointed out that even though the method is exactly conservative, it is not exact. Due to the uncoupled evolution of the edges, a certain amount of information is lost, resulting in approximation errors that increase with the size of Δt. Thus, there could be accuracy constraints on Δt that are more stringent than the stability constraints. Below, we outline the required additions to *particleclaw* (Sect. 4.1), describe the synchronization step (Sect. 4.2), and show how exact area balance is achieved (Sect. 4.3).

4.1 Virtual Domain, Excess Area, Virtual Area, and Area Credit

One key idea of the particle approach is that the numerical solution can be advanced on each edge, without considering the coupling with other edges at the network nodes. Clearly, this evolution incurs an error near the two edge boundaries, since the coupling information is not used. However, in Sect. 4.2 we design the coupling step such that the arising gaps in information near the edge boundaries can be filled (with satisfactory accuracy) during the coupling step.

The coupling-free evolution on an edge $x \in [0, L]$ works very similarly to the basic *particleclaw* methodology described in Sect. 3, with a special twist at the first and the last particle, as described below. When no particles share the same x-position, then all particles are simply moved according to the method of characteristics (11). Since this applies also to the first and the last particle, the domain of representation of the solution $[x^1(t), x^N(t)]$ does in general not match with the computational domain $[0, L]$. In the case that particles extend beyond the edge into the *virtual domain*, i.e. $x^1 < 0$ or $x^N > L$, the similarity interpolant (10) defines the numerical solution in particular up to $x = 0$ or $x = L$, respectively. In addition, it defines an *excess area* $\int_{x^1}^0 U_P(x)\,\mathrm{d}x$ or $\int_L^{x^N} U_P(x)\,\mathrm{d}x$, respectively. In the case that particles do not reach the edge boundary, i.e. $x^1 > 0$ or $x^N < L$, we can define an extrapolation on $[0, x^1]$ or $[x^N, L]$, respectively, if we are given a value for the *virtual area* $\int_0^{x^1} u\,\mathrm{d}x$ or $\int_{x^N}^L u\,\mathrm{d}x$, respectively. How to construct this extrapolation is described in Sect. 4.3.

The merging of colliding particles works as described in Sect. 3.2, with one exception. If for the merging of particles i and $i + 1$, there is no particle $i - 1$, then a new particle is added at $x^i - d$ with the same value as u^i, and a "left-sided" *area credit* $I_L = d\,u^i$ is recorded. Similarly, if there is no particle $i + 2$, a new particle is added at $x^{i+1} + d$ with the same value as u^{i+1}, and a "right-sided" *area credit* $I_R = d\,u^{i+1}$ is recorded. After these insertions, the merge can be performed.

4.2 Synchronization Step

The synchronization of area between the edges happens by moving area from edges in which information is flowing into the node to edges from which information is flowing out of the node. As introduced in Sect. 2.2, we denote the edges on which vehicles enter the node "ingoing". On the ingoing edge $i \in \{1, \dots, n\}$, the position of the node is at $x = L_i$. Conversely, the edges on which vehicles exit the node are called "outgoing". On the outgoing edge $j \in \{n + 1, \dots, n + m\}$, the position of the node is at $x = 0$. Since the LWR model (1) admits characteristic velocities of either sign, information can propagate either into or away from the node, both for ingoing and outgoing edges. Therefore, we introduce further terminology: edges for which information is going into the node are called "influencing", edges for which information is going away from the node are denoted "affected", and edges with zero characteristic velocity are called "neutral". When solving a generalized Riemann problem at a node, the procedure described in the last paragraph of Sect. 2.2 implies that all edges for which the flux changes (from $f_i(u_i)$ to γ_i) become automatically affected or neutral. Influencing edges arise if $\gamma_i = f_i(u_i)$ and in addition: $u_i < u_i^*$ for ingoing edges and $u_i > u_i^*$ for outgoing edges.

Let us now focus on one node in the network, with n ingoing and m outgoing edges. In the following, subscripts denote the edge index, and superscripts denote the particle index. With this notation, the last particle on the ingoing edge $i \in \{1, \dots, n\}$ is $(x_i^{N_i}, u_i^{N_i})$, and the first particle on the outgoing edge $j \in \{n + 1, \dots, n + m\}$ is (x_j^1, u_j^1). We design our approach such that at the end of the synchronization step (and thus at the beginning of the evolution on each edge) the first and last particle on each edge arise as solutions of the generalized Riemann problems described in Sect. 2.2. Consequently, due to (6), we have the m conditions

$$\sum_{i=1}^{n} a_{i,j} f_i(u_i^{N_i}) = f_j(u_j^1) \quad \forall j \in \{n+1,\ldots,n+m\}, \tag{15}$$

given by the desired destinations matrix A, and—if necessary—further $n - 1$ conditions

$$f_i(u_i^{N_i}) = \beta c_i \quad \forall i \in \{1,\ldots,n\} \tag{16}$$

for some $\beta \in \mathbb{R}$, given by the merging vector $c \in \mathbb{R}^n$.

We now derive, first for ingoing edges, an evolution equation for the excess area or virtual area (see Sect. 4.1) that is between the end of the edge L_i and the last particle $x_i^{N_i}$. Letting u denote a true solution of the hyperbolic conservation law (1), this area $I_i = \int_{L_i}^{x_i^{N_i}} u(x,t)\,dx$ has the rate of change

$$\frac{d}{dt} I_i = f_i(u_i(L_i,t)) - f_i(u_i^{N_i}) + u_i^{N_i} f_i'(u_i^{N_i}). \tag{17}$$

Here, the first two terms come from the PDE (1), and the third term stems from the fact that $x_i^{N_i}$ moves with velocity $f_i'(u_i^{N_i})$. Analogously, we compute the evolution of the excess/virtual area $I_j = \int_0^{x_j^0} u(x,t)\,dx$ on an outgoing edge as

$$\frac{d}{dt} I_j = f_j(u_j(0,t)) - f_j(u_j^1) + u_j^1 f_j'(u_j^1). \tag{18}$$

At the beginning of the coupling-free evolution, the first and last particle are on the edge boundaries, and thus all excess/virtual areas are zero. At the end of the coupling-free evolution, at the considered node, we know the excess area for all influencing edges. Moreover, for all neutral edges the excess area is zero. The idea is now to use the evolution Eqs. (17) and (18), together with the relations (15) and (16), to determine the unknown virtual areas at all affected edges. Any area credit that has been taken due to merges at the first or last particle one an edge (see Sect. 4.1) is simply subtracted from this balance. Below, we present this idea for three fundamental cases: a bottleneck, a bifurcation, and a confluence. Many practically relevant road networks can be constructed from these three types of nodes.

Bottleneck

A bottleneck is a sudden change of the road conditions, i.e. $n = 1$ and $m = 1$. Here, the feasibility domain (7) is $\Omega = [0,d_1] \cap [0,s_2]$, where $d_1 = f_1(\min\{u_1,u_1^*\})$

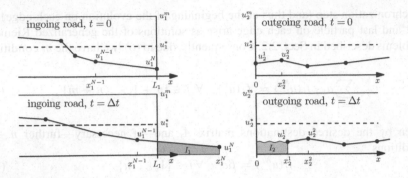

Fig. 1 Area tracking for a bottleneck. The excess area I_1 allows the computation of the virtual area I_2

and $s_2 = f_2(\max\{u_2, u_2^*\})$. Thus, the solution of the optimization problem (8) is $\gamma_1 = \min\{d_1, s_2\}$, which allows for two cases. If $\gamma_1 = d_1$, then all vehicles can pass through the node, and consequently edge 1 is influencing (or neutral), and edge 2 is affected (or neutral). Otherwise, if $\gamma_1 < d_1$, then not all vehicles can pass, and a jam is triggered. Consequently, edge 1 is affected, and edge 2 is influencing or neutral.

In either case, since the influx equals the outflux at all times, we have that $f_1(u_1(L_1, t)) = f_2(u_2(0, t))$ and $f_1(u_1^{N_1}) = f_2(u_2^1)$, and thus the evolution Eqs. (17) and (18) imply that

$$\frac{d}{dt}(I_2 - I_1) = u_2^1 f_2'(u_2^1) - u_1^{N_1} f_1'(u_1^{N_1}) = C,$$

which is a known time-independent quantity. Together with $I_1(t) = I_2(t) = 0$, we obtain a relation $I_2(t + \Delta t) - I_1(t + \Delta t) = C\Delta t$ that allows to obtain I_2 from I_1 (in the case $\gamma_1 = d_1$), or I_1 from I_2 (in the case $\gamma_1 < d_1$). Figure 1 shows an example of the movement of particles into the virtual domain, and the evolution of excess area (I_1) and virtual area (I_2).

Bifurcation

In a bifurcation, one road splits into two, i.e. $n = 1$ and $m = 2$. The relative numbers of vehicles that exit the node on edge 2 or edge 3 are given by $a_{1,2}$ and $a_{1,3} = 1 - a_{1,2}$, respectively. Here, the feasibility domain (7) is $\Omega = \{\gamma_1 \in [0, d_1] \mid a_{1,j}\gamma_1 \in [0, s_j] \; \forall \, j \in \{2, 3\}$, where $d_1 = f_1(\min\{u_1, u_1^*\})$ and $s_j = f_j(\max\{u_j, u_j^*\}) \; \forall \, j \in \{2, 3\}$. The solution of the optimization problem (8) is given by $\gamma_1 = \min\{d_1, \frac{s_2}{a_{1,2}}, \frac{s_3}{a_{1,3}}\}$. This allows two possibilities. If $\gamma_1 = d_1$, then all vehicles can pass through the node, and edge 1 is influencing (or neutral), and edges 2 and 3 are affected (or neutral). Otherwise, if $\gamma_1 < d_1$, then not all vehicles can pass, and edge 1 becomes affected. Of the outgoing edges, the one that causes the congestion is influencing or neutral, while the other outgoing edge is affected or neutral.

In either situation, we can use the evolution Eqs. (17) and (18) to determine the virtual areas, since (15) provides two conditions that relate I_1, I_2, and I_3 with each other. As an example, consider the case of edge 2 being influencing, and edges 1 and 3 affected. Here, I_2 is known (by integrating the interpolant (10)). From (17) and (18) we first compute $C = \frac{d}{dt}(I_1 - \frac{1}{a_{1,2}}I_2) = u_1^{N_1} f_1'(u_1^{N_1}) - \frac{1}{a_{1,2}}u_2^1 f_2'(u_2^1)$, and then obtain $I_1(t+\Delta t) = \frac{1}{a_{1,2}}I_2(t+\Delta t)+C\Delta t$. Knowing I_1, we can then determine I_3, by considering $\frac{d}{dt}(I_3 - a_{1,3}I_1)$.

Confluence

Here, two roads converge into one, i.e. $n = 2$ and $m = 1$. Flux conservation simply states that $\gamma_1 + \gamma_2 = \gamma_3$. The feasibility domain (7) is now two-dimensional: $\Omega = \{(\gamma_1, \gamma_2) \in [0, d_1] \times [0, d_2] \mid \gamma_1 + \gamma_2 \in [0, s_3]\}$, where $d_i = f_i(\min\{u_i, u_i^*\}) \ \forall \, i \in \{1, 2\}$ and $s_3 = f_3(\max\{u_3, u_3^*\})$. For the optimization problem (8), we distinguish two cases. If $d_1 + d_2 \le s_3$, then all vehicles can pass through the node, and the ingoing edges are influencing (or neutral), and the outgoing edge is affected (or neutral). In contrast, if $d_1 + d_2 > s_3$, then a jam is triggered. How the backwards going information distributes among the two ingoing edges depends now on the merging vector (c_1, c_2). The following outcomes are possible: (i) all vehicles from edge 1 get through, i.e. $\gamma_1 = d_1$ and a jam occurs on edge 2, i.e. $\gamma_2 < d_2$. In this case, edge 1 is influencing, edge 2 is affected, and edge 3 is influencing or neutral; (ii) a jam occurs on both ingoing edges, i.e. $\gamma_1 < d_1$ and $\gamma_2 < d_2$. Here, both ingoing edges are affected, and the outgoing edge is influencing or neutral; (iii) is the same as (i), but with edges 1 and 2 reversed.

Again, the virtual areas at the affected edges can be obtained systematically from the known virtual areas at the influencing and neutral edges. In the case that all vehicles pass through the node, we have $I_3(t + \Delta t) = I_1(t + \Delta t) + I_2(t + \Delta t) + \left(u_3^1 f_3'(u_3^1) - u_1^{N_1} f_1'(u_1^{N_1}) - u_2^{N_2} f_2'(u_2^{N_2})\right)\Delta t$. In turn, if jamming occurs, e.g. on both ingoing edges, we use the knowledge of the merging ratios $\frac{\gamma_1}{\gamma_2} = \frac{c_1}{c_2}$ to determine I_1 and I_2 from I_3. The methodology is analogous to the cases presented above.

4.3 Representation of Virtual Area by Particles

In Sect. 4.2, we have described how the virtual area at affected edges can be determined. The presented methodology tracks area exactly, and thus the numerical approach is exactly conservative. In order to finalize the synchronization step, we must account for the imbalance of area by changing parts of the solution, thereby creating or removing area.

Throughout the design of *particleclaw*, the two main guiding principles are: (i) conserve $\int u\,dx$ exactly and as locally as possible; and (ii) ensure that the approach is TVD, i.e. no spurious overshoots or undershoots are generated. We conclude the synchronization step by modifying the numerical solution, while following these two principles. The conservation principle means that the excess area and the virtual area must be converted into actual area under particles on the edges near where that area was recorded. The locality requirement is respected by a finite domain of influence: the solution is never modified further than Δt times the maximal particle velocity from the node. The TVD condition means that the modified solution does not possess values above or below the range of values that is presented in the nearby particles and the solution $\hat{u}$ that comes from the generalized Riemann problem (8). The implementation outlined below violates the TVD condition only under exceptional circumstances. In fact, in the numerical experiments presented in Sect. 5, this non-TVD "last resort" solution turned out to never be required.

The precise methodology is as follows. First, the numerical solution is truncated or extrapolated to the edges, by adding a particle either on the interpolant, or with the same value as the closest particle, respectively. All excess areas (i.e. the parts that extend beyond the domain of each edge) can now be removed, since their contribution has been accounted for in the step presented in Sect. 4.2. Now, the generalized Riemann solver is called with the new values on the node position. The resulting $\hat{u}$-values are added to the solution as new extremal particles (unless the value has not changed, $\hat{u} = u$, in which case no particle is added) with the same x-value as the extremal particle. Being able to have two particles with the same x-value is a peculiarity of the particle method, and it implies that the particle insertion does not change the area under the solution.

With this, on each edge we have a numerical solution, whose extremal states are solutions of the generalized Riemann problems. What is left is the accounting for the correct balance of area. We do so by modifying the numerical solution, however, without changing the previously found new extremal states. This guarantees that at the end of the synchronization step, the fluxes at the all the nodes are solutions of the generalized Riemann problem. In the preceding subsections we have derived the area I that is supposed to be under the solution between the network node and the extremal particle. In general, the area under the numerical solution constructed above by simple interpolation/extrapolation has a small disparity with I. The difference, ΔI, needs to be added to the affected edge, by modifying the numerical solution. Without loss of generality let us examine an *outgoing* affected edge. We look for a new solution to replace the current solution near the end of the edge. This new solution must have an area difference ΔI from the area of the current solution, and the first particle must remain unchanged in order to satisfy the generalized Riemann problem. We follow the steps below (starting with $k = 2$) to represent the required change in area using particles (illustrated in Fig. 2):

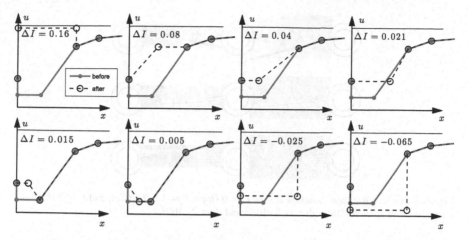

Fig. 2 Reconstructed solution for various ΔI. The *thick gray line* is the original solution, and the *dashed black line* the "correct area" solution that replaces it, whose area is greater by ΔI

1. Attempt a solution that comprises a constant part and a linear part so that it connects particles 1 and k in a continuous fashion. If such a solution can provide ΔI, accept it.
2. Attempt a solution that has a constant value u in the domain $[x_1, x_k]$, where $u \in [\min_{i \leq k}(u_i), \max_{i \leq k}(u_i)]$. If such a solution can provide ΔI, accept it. If this solution is not accepted, and $x_k < \Delta t \max_u(f'(u))$, increase k by one and go back to (1), otherwise, go to (3).
3. Use a solution that has a constant value u in the domain $[x_1, x_k]$ with the value of u needed to match ΔI.

In the algorithm above, step (3) is a fail-safe mechanism that will always result in a solution, but the solution may be non-TVD or even unphysical.

Figure 2 shows eight exemplary cases using the same initial particle configuration but different ΔI. In these examples, k could not increase beyond 4: $\Delta I = 0.16$ resulted in a solution at step (3) with $k = 4$; $\Delta I = 0.08, 0.04, 0.021$, with step (1) and $k = 4$; $\Delta I = 0.015, 0.005$ with step (1) and $k = 3$; $\Delta I = -0.025$ with step (2) and $k = 4$; and $\Delta I = -0.065$ with step (3) and $k = 4$.

Performing this area reconstruction on all affected edges guarantees that area is preserved exactly; it may increase or decrease the number of particles.

5 Numerical Results

5.1 Bottleneck Test Case

As a first test case, we consider the simplest non-trivial network: a bottleneck that consists of two edges, each of length $L_1 = L_2 = 1$ that are joined linearly, as shown

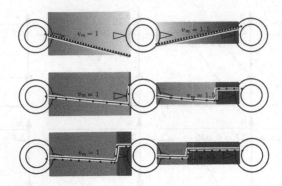

Fig. 3 Bottleneck test case. Solution at time $t = 0$ (*top*), $t = 1.5$ (*middle*), and $t = 3$ (*bottom*), computed with *particleclaw* with $h = 8 \cdot 10^{-2}$ and $d = 2 \cdot 10^{-2}$

in Fig. 3. In dimensionless variables, the maximum traffic densities on the two roads are $u_1^m = 2$ and $u_2^m = 1$, and the maximum vehicle velocities are $v_1^m = 1$ and $v_2^m = 1.5$. This example can be interpreted as a model for the situation of two lanes merging into one lane, and at the same position the speed limit increasing by 50 %. The initial conditions are $u_1(x) = 1 - x$ and $u_2(x) = 0.8\,x$, and the boundary conditions of $u_1(0,t) = 1$ and $u_2(1,t) = 0.8$ are prescribed whenever information is entering the edge through these boundaries. The final time is $t_{\text{final}} = 3$. In Fig. 3 (and also in Fig. 5), the solution is shown in two ways: the shade of gray becomes darker with higher values of u, and the plot of u vs. x is overlaid with dots representing the particles of the method. Each edge is annotated with the maximum velocity v_m and two arrows showing the direction of vehicles. The thickness of each segment is proportional to its maximum density u_m.

It is shown in [11] that *particleclaw* itself is second-order accurate with respect to the maximum distance of particles (away from shocks or when using shock sharpening). What we investigate here is the error as a function of the size of the synchronization time step Δt, while having a large number of particles, such that the error due to the spacial approximation is negligible. Specifically, we provide the following two parameters for *particleclaw*: an initial/desired distance of particles of $h = 8 \cdot 10^{-5}$ is given, and the distance of particles that are inserted near a shock is $d = 2 \cdot 10^{-5}$. Both parameters result in a very fine resolution, using more than 10,000 particles on each edge. We choose the time step from $\Delta t = 2^{-k/2}$, where $k \in \{4, \ldots, 15\}$, and compute a reference solution using $k = 20$. Since here we do not wish to measure the error at a shock, we evaluate the error in the segment $x \in [0, 0.3]$ on edge 2, in which the solution is smooth, but non-linear. Figure 4 shows the $L^{\infty}([0, 0.3])$ and $L^2([0, 0.3])$ errors of the approximate solution as a function of the synchronization time step, Δt. One can see a second-order dependence, i.e. the error scales like $O((\Delta t)^2)$.

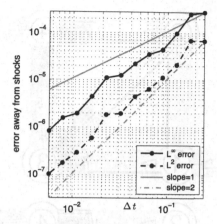

Fig. 4 Bottleneck test case. The error as a function of the synchronization time step Δt exhibits second order convergence behavior

5.2 Simulation of a Diamond Network

An important point of *particleclaw* is that it tends to yield high quality numerical solutions (including shocks) for very few particles (see e.g. [11, 13]). Here, we demonstrate this feature for a so-called diamond network. This network of seven edges and six nodes is shown in Fig. 5. Vehicles enter in the node marked "A" driving towards "1"; then the traffic flow splits into a N and a E direction. At "2", drivers again decide to go SE or straight E. At "3" and "4", two confluences happen, and finally vehicles exit the network at "B". We choose all roads identical $u_1^m = 1$ and $v_1^m = 1$, and at all network nodes, an even split of traffic flow occurs. The lengths of all the roads is taken to be 1 (even if it does not seem so in the plot). The initial conditions are also identical on all the roads: $u(x, 0) = 0.4 + 0.4 \cos(3\pi x)$. We evolve the solution until the time $t_{\text{final}} = 2$. Figure 5 shows the computational results when applying *particleclaw*, with three different types of resolutions (in space and time). The top solution has an initial/desired distance of particles of $h = 2 \cdot 10^{-2}$, the distance of particles that are inserted near a shock is $d = 5 \cdot 10^{-3}$, and the synchronization time step is $\Delta t = 1 \cdot 10^{-2}$. In the middle solution, all of these numbers are multiplied by a factor of 5, and in the bottom solution by a factor of 20. That means that the top possesses 20 times the resolution than the bottom, and it is therefore about 400 times as costly to compute. One can observe that the higher resolution at the top yields sharper features. However, the presence and position of the shocks (and other features) is well captured by the resolution in the middle, and even at the bottom most features are visible quite clearly.

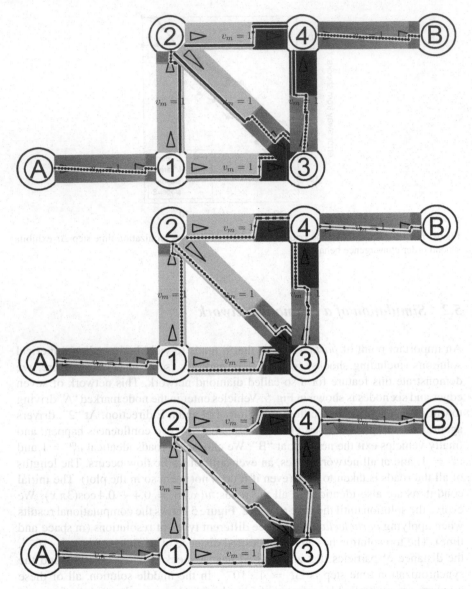

Fig. 5 Diamond network test case with various resolutions. The solutions have parameters $h = 2 \cdot 10^{-2}s$, $d = 5 \cdot 10^{-3}s$, and $\Delta t = 1 \cdot 10^{-2}s$, where $s = 1$ at the *top*, $s = 5$ in the *middle*, and $s = 20$ at the *bottom*

6   Conclusions and Outlook

The framework presented in this paper demonstrates that the characteristic particle method *particleclaw* can be generalized to solving first order traffic models on road networks in a robust, accurate, and exactly conservative fashion. The presented methodology allows to apply the basic one-dimensional solver on each edge. The coupling of traffic states at network nodes is achieved by a special synchronization step that ensures exact conservation properties by applying a proper transfer of area between edges. A fundamental ingredient in this synchronization is the invoking of generalized Riemann solvers that have been developed for traffic networks. In this paper, a specific focus lies on 1-to-1, 1-to-2, and 2-to-1 highway junctions. However, the presented methodology applies to more general types of networks, as long as the evolution on each edge is described by a scalar hyperbolic conservation law, and the problem description allows to formulate coupling conditions at the network nodes.

As previously shown [11], *particleclaw* itself can be made second-order accurate with respect to the spacing between particles. The numerical examples investigated here confirm that the presented method is also second order accurate with respect to the time step Δt between synchronization events. Moreover, it is demonstrated that the presented approach yields good quality numerical solutions (with an accurate location of shocks), even when using only few particles on each edge.

In contrast to many traditional numerical methods, here the synchronization time step Δt is not limited by the spacing between particles (i.e. no CFL stability restriction). It is only limited by the length of the edges, insofar as between two synchronization events, information is not allowed to travel beyond a single edge. In general, this allows for much larger values for Δt than traditional methods do. An exception is posed by networks that possess a few very short edges. These can be treated significantly more efficiently by a simple generalization of the presented approach, namely by using a different Δt for each network node, depending on the shortest edge connected to it. Thus, edges that neighbor short edges would need to synchronize more often, but the rest of the network would not.

The accurate location of shocks with only few particles on each edge, and the possibility of invoking the coupling conditions at network nodes relatively rarely, allow for fast and memory-efficient implementations. This makes the approach attractive for the simulation of large road networks, as well as the optimization, design, and control of the traffic flow on such large networks. It is also conceivable that the presented methodology can be adapted to other kinds of nonlinear flows on networks, for instance continuum models of supply chains [1, 20].

Acknowledgements Y. Farjoun was partially financed by the Spanish Ministry of Science and Innovation under grant FIS2008-04921-C02-01. The authors would like to acknowledge support by the National Science Foundation. B. Seibold was supported through grant DMS–1007899. In addition, B. Seibold wishes to acknowledge partial support by the National Science Foundation through grants DMS–0813648 and DMS–1115278. Y. Farjoun wishes to acknowledge partial support through grant DMS–0703937. In addition, Farjoun thanks Luis Bonilla at the UC3M and Rodolfo R. Rosales at MIT for providing a framework under which this work was done.

References

1. D. Armbruster, D. Marthaler, C. Ringhofer, Kinetic and fluid model hierarchies for supply chains. Multiscale Model. Simul. **2**, 43–61 (2004)
2. A. Aw, M. Rascle, Resurrection of second order models of traffic flow. SIAM J. Appl. Math. **60**, 916–944 (2000)
3. M.J. Berger, J. Oliger, Adaptive mesh refinement for hyperbolic partial differential equations. J. Comput. Phys. **53**, 484–512 (1984)
4. G. Bretti, R. Natalini, B. Piccoli, Fast algorithms for the approximation of a traffic flow model on networks. Discret. Contin. Dyn. Syst. Ser. B **6**, 427–448 (2006)
5. G. Bretti, R. Natalini, B. Piccoli, Numerical approximations of a traffic flow model on networks. Netw. Heterog. Media **1**, 57–84 (2006)
6. B. Cockburn, C.-W. Shu, The local discontinuous Galerkin method for time-dependent convection-diffusion systems. SIAM J. Numer. Anal. **35**, 2440–2463 (1988)
7. G.M. Coclite, M. Garavello, B. Piccoli, Traffic flow on a road network. SIAM J. Math. Anal. **36**, 1862–1886 (2005)
8. R. Colombo, Hyperbolic phase transitions in traffic flow. SIAM J. Appl. Math. **63**, 708–721 (2003)
9. L.C. Evans, *Partial Differential Equations*. Graduate Studies in Mathematics, vol. 19 (American Mathematical Society, Providence, 1998)
10. Y. Farjoun, B. Seibold, Solving one dimensional scalar conservation laws by particle management, in *Meshfree Methods for Partial Differential Equations IV*, ed. by M. Griebel, M.A. Schweitzer. Lecture Notes in Computational Science and Engineering, vol. 65 (Springer, Berlin, 2008), pp. 95–109
11. Y. Farjoun, B. Seibold, An exactly conservative particle method for one dimensional scalar conservation laws. J. Comput. Phys. **228**, 5298–5315 (2009)
12. Y. Farjoun, B. Seibold, A rarefaction-tracking method for conservation laws. J. Eng. Math. **66**, 237–251 (2010)
13. Y. Farjoun, B. Seibold, An exact particle method for scalar conservation laws and its application to stiff reaction kinetics, in *Meshfree Methods for Partial Differential Equations V*, M. Griebel, M.A. Schweitzer. Lecture Notes in Computational Science and Engineering, vol. 79 (Springer, Heidelberg/Berlin, 2011), pp. 105–124
14. M.R. Flynn, A.R. Kasimov, J.-C. Nave, R.R. Rosales, B. Seibold, Self-sustained nonlinear waves in traffic flow. Phys. Rev. E **79**, 056113 (2009)
15. S.K. Godunov, A difference scheme for the numerical computation of a discontinuous solution of the hydrodynamic equations. Math. Sb. **47**, 271–306 (1959)
16. B.D. Greenshields, A study of traffic capacity. Proc. Highw. Res. Rec. **14**, 448–477 (1935)
17. A. Harten, B. Engquist, S. Osher, S. Chakravarthy, Uniformly high order accurate essentially non-oscillatory schemes, III. J. Comput. Phys. **71**, 231–303 (1987)
18. D. Helbing, Traffic and related self-driven many-particle systems. Rev. Mod. Phys. **73**, 1067–1141 (2001)
19. M. Herty, A. Klar, Modelling, simulation and optimization of traffic flow networks. SIAM J. Sci. Comput. **25**, 1066–1087 (2003)
20. M. Herty, A. Klar, B. Piccoli, Existence of solutions for supply chain models based on partial differential equations. SIAM J. Math. Anal. **39**, 160–173 (2007)
21. H. Holden, L. Holden, R. Hegh-Krohn, A numerical method for first order nonlinear scalar conservation laws in one dimension. Comput. Math. Appl. **15**, 595–602 (1988)
22. H. Holden, N.H. Risebro, A mathematical model of traffic flow on a network of unidirectional roads. SIAM J. Math. Anal. **26**, 999–1017 (1995)
23. H. Holden, N.H. Risebro, *Front Tracking for Hyperbolic Conservation Laws* (Springer, New York, 2002)

24. P.D. Lax, B. Wendroff, Systems of conservation laws. Commun. Pure Appl. Math. **13**, 217–237 (1960)
25. M.J. Lighthill, G.B. Whitham, *On kinematic waves. II. A theory of traffic flow on long crowded roads*. Proceedings of Royal Society A, Piccadilly, London, vol. 229 (1955), pp. 317–345
26. X.-D. Liu, S. Osher, T. Chan, Weighted essentially non-oscillatory schemes. J. Comput. Phys. **115**, 200–212 (1994)
27. H.J. Payne, FREEFLO: A macroscopic simulation model of freeway traffic. Transp. Res. Rec. **722**, 68–77 (1979)
28. W.H. Reed, T.R. Hill, *Triangular Mesh Methods for the Neutron Transport Equation*, Tech. Rep. LA-UR-73-479, Los Alamos Scientific Laboratory (1973)
29. P.I. Richards, Shock waves on the highway, Oper. Res. **4**, 42–51 (1956)
30. B. Seibold, *Particleclaw*. Website. http://www.math.temple.edu/~seibold/research/particleclaw
31. B. van Leer, Towards the ultimate conservative difference scheme II. Monotonicity and conservation combined in a second order scheme. J. Comput. Phys. **14**, 361–370 (1974)
32. H.M. Zhang, A theory of non-equilibrium traffic flow. Transp. Res. B **32**, 485–498 (1998)

Meshfree Modeling in Laminated Composites

Daniel C. Simkins, Jr., Nathan Collier, and Joseph B. Alford

Abstract A problem of increasing importance in the aerospace industry is in detailed modeling of explicit fracture in laminated composite materials. For design applications, the simulation must be capable of initiation and propagation of changes in the problem domain. Further, these changes must be able to be incorporated within a design-scale simulation. The use of a visibility condition, coupled with the local and dynamic nature of meshfree shape function construction allows one to initiate and explicitly open and propagate holes inside a previously continuous problem domain. The method to be presented naturally couples to a hierarchical multi-scale material model incorporating external knowledege bases to achieve the goal of a practical explicit fracture modeling capability for full-scale problems.

Keywords Meshfree modeling • laminated composites • visibility condition

1 Introduction

Physical phenomena in some important areas, such as fracture, fundamentally occur in a discontinuous fashion. On the other hand, most methods for solving structural mechanics problems, e.g. the finite element method, are formulated to

D.C. Simkins, Jr. (✉) · J.B. Alford
Department of Civil and Environmental Engineering, University of South Florida, 4202 E. Fowler Ave. ENB118, Tampa, FL, 33620, USA
e-mail: dsimkins@usf.edu

N. Collier
Applied Mathematics & Computational Science and Earth Science & Engineering King Abdullah University of Science and Technology (KAUST), 4700 King Abdullah University of Science and Technology, Building UN1550-4, Office # 4332, Thuwal, 23955-6900, Kingdom of Saudi Arabia

M. Griebel and M.A. Schweitzer (eds.), *Meshfree Methods for Partial Differential Equations VI*, Lecture Notes in Computational Science and Engineering 89, DOI 10.1007/978-3-642-32979-1_14, © Springer-Verlag Berlin Heidelberg 2013

solve for continuous field variables. To model a crack in a continuum model requires embedding some model of discontinuity into the continuous formulation. Numerous approaches using this methodology have been introduced, such as X-FEM, cohesive cracks, element erosion, etc. These methods have achieved some degree of success. Purely discrete models, such as molecular dynamics (MD), can naturally evolve to open up discontinuities in a material. However, the number of molecules required for MD simulations capable of modeling practical engineering fracture problems is prohibitively large. Thus, there seems to be a place for a simple method to solve the continuum equations, but easily incorporate discontinuities. For laminated composites, the fiber orientation has a profound effect on how cracks can propagate. In mesh-based analysis methods, the mesh orientation may influence the crack direction and lead to a conflict with the physical requirements dictated by the fiber orientation. This paper presents such a method based on one of the so-called meshfree methods, in this case the Reproducing Kernel Particle Method (RKPM). The method explicitly models cracks and changing topology and is formulated to reduce or eliminate discretization choices from influencing the solution. Further, the model naturally provides a mechanism for incorporating information that is not contained in the governing continuum equations to determine crack initiation and direction. The present paper provides a deeper examination of the general approach in [6].

2 RKPM And Crack Modeling

The Reproducing Kernel Particle Method was first introduced in [5] and provides a method to construct a function basis for use in Galerkin solutions to partial differential equations without using a mesh, but rather through locally and perhaps dynamically determined interactions between particles. Since a mesh is not required to form the function space, it can evolve dynamically, and in particular, can be adjusted as the material topology changes. The RKPM method constructs a local interpolation field centered at each particle that is based on local interaction with neighboring particles. By introducing a visibility condition during a calculation, one can selectively limit the interaction between particles to effectively cut the material. The concept of a visibility condition originated with the work of Belytschko and co-workers, [1, 2].

2.1 Crack Morphology

A crack morphology and propagation algorithm that conserves mass and energy was developed for two-dimensions in [7]. This paper presents the extension of this algorithm to three-dimensional through-cracks of directed surfaces, a method the first author has dubbed the Particle Splitting Crack Algorithm (PSCA).

Fig. 1 Crack filament

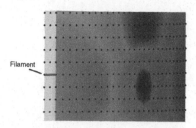

Filament

The essential concept is to identify a set of particles through the thickness that act together to form the crack tip. This set of particles is denoted a *filament* and the two-dimensional crack algorithm is then applied to filaments. Figure 1 depicts a set of particles and highlights a filament. The black dots are the particles, and the heavy red line identifies the set of particles that are identified as a filament, in this case there are three. Note that the use of the filament concept necessitates a self-similar particle discretization through the thickness of the shell. A scalar damage measure, to be discussed in Sect. 3.1, is computed for each particle in the domain. Then, for each filament, the member particle damage measures are used to compute an over all damage level for the filament. Many choices for this average are available, for example a simple maximum, or, in the present case, a simple average. If the filament damage level reaches or exceeds the failure threshold, as dicated by the material model, then the filament is split and the crack propagated. Propagation begins by searching the other filaments in the radius of support of the failed filament and locates the filament with the highest damage level. It is assumed that this filament will become the next crack tip. The failed filament is then split by splitting each particle in the filament. Once particles have split, the visibility condition is enforced by inserting a *crack panel* between the old tip and new tip and used to determine the updated visibility of particles. To help visualize the process, consider the example of a pressurized cylinder loaded to rupture. This calculation was performed using a total Lagrangian approach with coupled thermo-mechanical behavior. In Fig. 2 is the reference configuration with adjacent crack sides colored red and blue. Note that the mesh seen in this and subsequent figures is the mesh used for visualization, the computation is mesh free. The deformed configuration with explicitly opened cracks is shown in Fig. 3. Crack panels live in the reference configuration, and the sequence of panels for this particular problem are shown in Fig. 4.

Depending upon the geometry of the problem, a crack panel may not necessarily be planar. In the cylinder example, the crack panels may be planar, for orientations that are parallel to the radial direction or orthogonal to the axial direction. On the other hand, for panels lying on the concentric cylindrical surfaces, the panels are cylindrical. Consider a crack that may propagate at an angle with respect to the horizontal, as shown in Figs. 5–7. In this case, the panel is neither planar nor a simple geometric shape requiring special treatment.

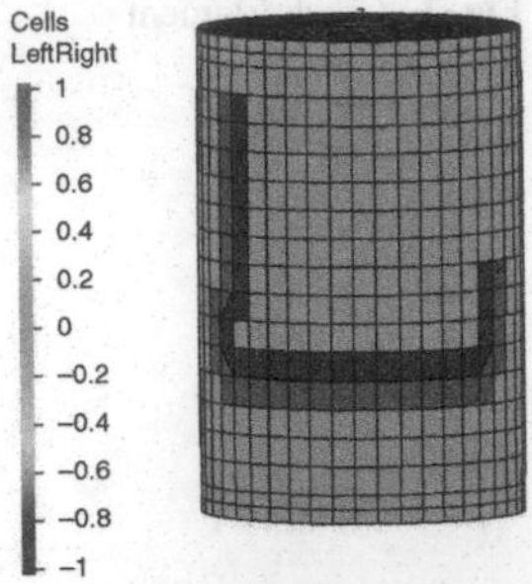

Fig. 2 Reference
configuration showing crack
path between the *red* and *blue*
colored regions. The cells
shown are the from the
visualization mesh

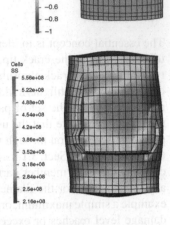

Fig. 3 Deformed
configuration with large crack
opening

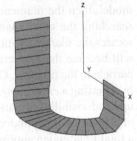

Fig. 4 Crack panels used to
enforce the visibility
condition located along the
crack *center line*

2.2 Visibility Condition

The visibility condition is enforced when determining connectivity of the particles.
The connectivity between particles and integration points to particles is determined
in the usual fashion. Then, each connectivity is checked to make sure the line of
sight (LOS) between the particle and target object are not blocked by a crack panel,
as determined by a line segment – crack panel intersection check. If the LOS is
blocked, the connectivity is severed, otherwise it is left intact. It is in the LOS –
crack panel intersection that detailed geometry of the crack panel may be necessary.
On the other hand, under refinement, a simple bi-linear representation of any crack
panel may be sufficient.

Fig. 5 Cylinder

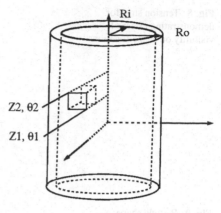

Fig. 6 Cylindrical box with *diagonal* crack

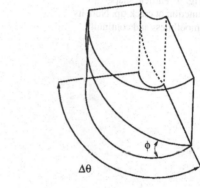

Fig. 7 Non-planar crack panel

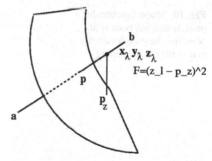

Imposition of the visibility condition introduces a sharp discontinuity on the shape functions of particles that have been split, effectively cutting the material. In Fig. 8 shows a tensions trip in which a crack is allowed to grow across the plate, then is arrested. The color contours depict the second dilitational strain invariant. Plots of the shape function at a crack tip and one that has been split are shown in Figs. 9 and 10, respectively. Finally, Fig. 11 shows the shape of the crack tip. In this figure, the larger red dots are the computational particles used to solve the PDE. The smaller white dots are are a much refined set of points where solution variables

Fig. 8 Tension strip for demonstrating effect of visibility on shape functions

Fig. 9 Particle shape function at crack tip. Note its smoothness and continuity

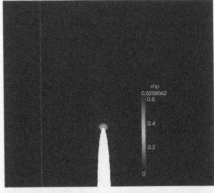

Fig. 10 Shape function for a particle that has been split. Note the sharp discontinuity across the crack opening

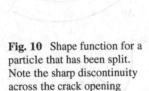

and position points are interpolated. Due to the large amount of data involved, the very fine interpolation points are restricted to a vicinity of the crack tip, and is evident along the bottom of the figure where no white colored points exist, only the red computational points. As can be seen, the tip is not very sharp, and may serve to mitigate the so-called crack tip singularity known in linear elastic fracture mechanics.

Fig. 11 Crack tip geometric shape highlighted in *white*. The *red* particles are the computational particles

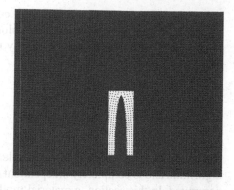

Fig. 12 Locally heated pressure cylinder with effective stress contours. The *blue* spots (at the intersection of the X shaped bands) are spots with an intense applied heat source. The stress is lower due to the thermal softening of the material

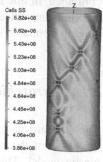

Fig. 13 Enlargement of the crack tip during propagation. The *black dots* show the computational particles

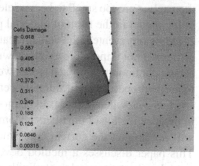

The particle splitting algorithm uses information determined from the constitutive model to determine when a particle should be split. Based upon this damage measure the crack propagation algorithm automatically chooses crack direction and propagation. The ability to couple particle splitting to a damage measure is immensely powerful. In fact, it can be viewed as a link between the structural mechanics equations and an externally (to the mechanics equations) determined quantity. There is no need for the external quantity to be continuous and its insertion into the simulation is done for each particle, and is thus discontinuous. Proper selection of the external damage measure allows a researcher to incorporate information from knowledge bases, or detailed physics-based multi-scale calculations. An example of the three-dimensional algorithm applied to a pressurized cylinder with localized heating is shown in Fig. 12, and a close up of a crack tip in Fig. 13.

The black dots represent particles, the color contours are of the hoop stress and external damage, respectively. The color contours in the figures are based on interpolated values from the particles by the graphics program used to display the images.

3 Driving The Algorithm

The PCSA described in Sect. 2 is dependent on a scalar damage measure to drive the algorithm, and is completely independent of choice of damage measure, or how it is computed. In this section, a material model suitable for laminated composites will be briefly introduced, then some specializations of the basic PCSA will be described making it suitable for laminated composites.

3.1 Hierarchical Multiscale Modeling

The standard method of modeling composites is the continuum approach where the effect of the composite fibers are idealized as material properties of a linear elastic model. This, in a sense, is a form of the multiscale problem. The fibers themselves are too small to be modeled explicitly and so this information is included in the model in the form of material properties. Even if the small scale of the fibers could be represented, the actual location of these in the matrix is knowledge that is difficult if not impossible to obtain with any reliable amount of certainty. The linear-elastic continuum approach works well for applications provided that the composite material properties are accurately determined.

However, when failure is to be modeled, especially cracking, the information lost in the idealization of the fibers becomes critical. Cracking models based on continua will not predict the correct crack morphology due to the absence of actual fibers. This paper discusses a method developed that returns physical knowledge to the model without the need to explicitly model fibers in the composite. This enables the straightforward use of continuum mechanics to solve crack propagation problems in laminate fibrous composites which will model the physical behavior.

In application to glassy polymer composites, the appropriate measure of damage lies in strain space. To properly account for the local effects of the fibers, fiber-matrix interface and the internal thermally induced strain, the states of strain are micro-mechanically enhanced using the methods described in [3,4,8]. This enhancement is performed at each evaluation point and operates on the homogeneous strain, i.e. it is external to the governing equations. The micro-mechanically enhanced strains are used to compute the strain invariants and are compared to critical values for these quantities.

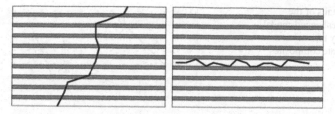

Fig. 14 Schematic of composite showing fibers in *gray* and matrix in *white*. The case on the *left* shows an unphysical crack morphology. The case on the *right* is permitted as no fibers are severed

3.2 External Knowledge Enhancement

The particle splitting algorithm used to model the crack propagation also suffers from the lack of explicitly modeled fibers. The morphology is only guided by the scalar measure of material utilization, therefore, there may be crack trajectories generated by this algorithm that are unphysical. Micro-mechanics and micro-mechanical enhancement effectively communicate the small scale information to the large scale. The determination of material failure, and whether it is matrix or fiber, is returned to the continuum solution. However, to avoid unphysical crack paths one must incorporate the knowledge of whether the failure is in the matrix or fiber phase. For example, at the micro-scale, it is known whether the matrix or fiber failed. If the matrix failed, than the crack cannot cross the fiber direction. The nodal search algorithm described in Sect. 2 does not account for such restrictions, since the macro-scale does not know about the micro-scale structure. To remedy this, a discrimination on the selection of possible new crack tips is applied to remove unphysical cases, such as cracks crossing fibers that have not failed. Fiber failure typically does not occur under normal loading. This situation can be seen in Fig. 14. On the left the fibers are oriented from left to right and the crack passes through them. This kind of morphology is restricted because it is unphysical. The material model returns to the crack algorithm information on whether the fiber failed or not. If the fiber has not failed, then the left picture in the figure indicates an unphysical crack path. The list of new potential crack tip particles are filtered to remove candidates that would lead to such a situation. The remaining potential new crack tips may lead to a crack path shown on the right in Fig. 14. Note that this method is dependent on the particle distribution being consistent with the local fiber orientation. While, in principle, a crack can move in any direction, it must have a particle available to become the new tip. Theroetically, one might expect this to work in an average sense, that is the crack might deviate from the physically allowable direction slightly, but asymptotically be correct since the crack is not restricted to follow any particular path. This does not happen in practice, since the mechanical fields are localized near the crack tip and will always, or nearly always, be biased to the nearest particles, and hence their relative distribution. A current research topic is to alleviate this restriction. In a multi-layer laminate, cracking does not propagate

Fig. 15 Details of the
knowledge-enhanced model
as they apply to
multi-laminate composites

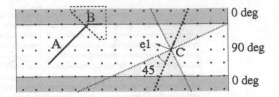

from one laminate to another. Instead, delamination occurs when the crack reaches
the interface. This is handled by limiting the meshfree particles to which a crack may
propagate. Consider Fig. 15 where a portion of a 0°/90°/0° laminate composite is
shown. The 0° plies are shaded and the fibers oriented left to right. The 90° ply is
left white and its fibers are aligned coming in and out of the page. The small circles
drawn represent the distribution of meshfree particles.

In this example, a crack has initiated at a particle labeled A. The crack is shown
by the dark, solid line connected to particle A. The crack is free to propagate in
the lamina until it reaches the interface, represented here by a particle B. The
normal cracking algorithm allows propagation to any neighboring particle in front
of particle B. In this case these candidate particles are indicated by a region defined
by a dashed line. However, if at a laminate interface, the propagation cannot move
into the next lamina, nor return into the original lamina. Once a crack reaches this
interface, the crack may only propagate in such a way as to cause delamination. This
is consistent with what is physically seen in experiments.

When cracks initiate inside a lamina, a criteria also governs the subset of particles
to which cracks may propagate. Consider a particle labeled C again in Fig. 15 which
represents the initiation of a crack. To determine candidate propagation particles, we
use the following procedure.

1. Extract the state of strain in the plane normal to the fiber direction from the total
   strain tensor. For the 90° ply in Fig. 15 this corresponds to the state of strain in
   the plane corresponding to that of the page.
2. Determine the maximum principle strain (ϵ_{max}) and accompanying direction. In
   Fig. 15 this direction is depicted as a vector originating at particle C and labeled,
   ϵ_{max}. This is observed to be the direction in which cracks open in the laminates.
3. The crack propagation direction is perpendicular to the maximum principle
   strain. Due to coarse granularity of the particle distribution, it is unlikely that
   a particle to which a crack may propagate exists in this direction. Candidate
   particles are chosen to lie within the 45° cone around this direction as well as in
   the current layer of the lamina. These particles are shown with a light shading in
   Fig. 15.

These candidate particles are determined as a crack initiates, and maintained as it
propagates. As the crack propagates and searches for new crack directions, particles
which are not in the candidate subset determined when the crack initiated are
ignored.

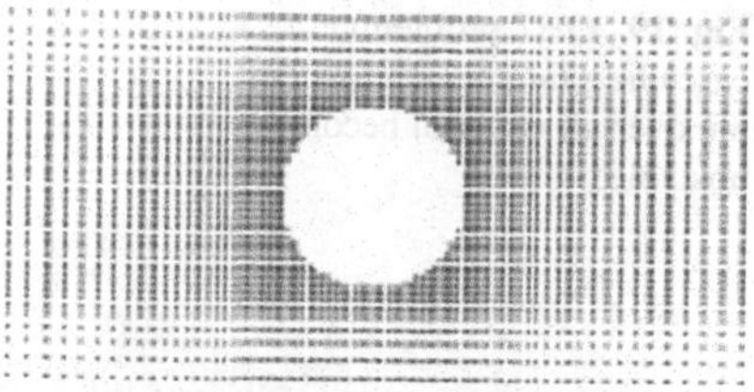

Fig. 16 Particle distribution
for OHT problem

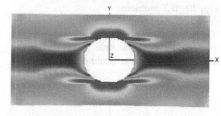

Fig. 17 OHT problem: onset
of failure in the epoxy matrix

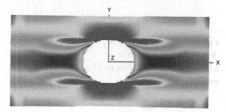

Fig. 18 OHT problem:
horizontal propagating cracks
parallel to the local fiber
direction, consistent with a
matrix failure in the absence
of fiber fracture

4 Results

The efficacy of the method is demonstrated on two classic problems in laminated composites, the open hole tension (OHT) specimens and the inter-laminar tension (ILT) specimen. The material model used is that described in [3], however, the specific material parameters are specific to a proprietary material formulation and can not be presented here.

The open-hole tension specimen is discretized as shown in Fig. 16. The fibers are oriented horizontally, i.e. parallel to the long edge and the entire plate is stretched along its length by applied forces along the left and right edges. In an isotropic material, a crack would be expected to form at the top and bottom of the hole and propagate vertically. However, in this case, the fibers do not fail, and thus this mode of failure is prohibited. Figures 17–19 show three snapshots during the loading sequence. The area in red is the region of material that is at, or near the failure limit. It is seen that the top and bottom of the hole do show the expected pattern, but under continued loading, the region spreads horizontally. This is due to the fibers not failing, and in Fig. 18 one can begin to see horizontal cracks opening up. In the final picture, a cascading failure event has occurred with many horizontal cracks developing. The inter-laminar tension specimen consists of an arched beam with multiple plies. In this example, the layup is such that the fibers along the inside radius are parallel to the circumference, then perpendicular out of the page, again

Fig. 19 OHT problem:
cascading failure event when
the over all material becomes
unstable

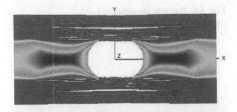

Fig. 20 ILT problem
definition and layup. *Blue* is a
0° ply, *red* a 90° ply. The
grey outline shows the final
displaced shape under loading

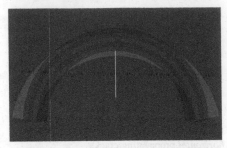

Fig. 21 ILT problem:
delamination initiation
followed by development of
intraply cracks

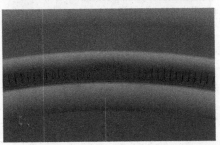

parallel to the circumference, etc. There are five plies in all. The bottoms of the arch
are pulled outward, like a wish bone. Figure 20 shows the model and layup, the gray
outline shows the final deformed shape.

Typically in laminated composites, cracks first develop within a ply, and then
either propagate within the ply, or arrest at an inter-laminar boundary. Under some
cases, the crack can lead to a delamination. Figure 21 shows that the inner ply does
not crack at all, rather, the next ply develops intra-ply cracks and delamination
occurs between the inner-most ply and the next one. This is due to the fibers in
the inner ply being parallel to the circumference, so that ply cannot crack under this
loading.

5 Conclusions

The use of the knowledge-enhanced meshfree model was successful in producing
an analysis which correctly predicted the location and propagation of failure
without a priori knowledge of where failure should occur. This in and of itself
is a significant result. To accomplish this, a linear-elastic model was used with

strain-based discriminators of failure. The states of strain were mirco-mechanically enhanced to correct for the residual matrix-fiber strains present. Finally the meshfree crack propagation algorithm was enhanced with physical knowledge, the result of which is a high fidelity solution.

While the results are interesting and faithfully reproduce test results, the mathematical basis for the solutions to the governing differential equations is unknown. The basic method satisfies the standard criterion for Galerkin solutions of PDE's, but in this case, the underlying topology of the problem domain is dynamic. So, it is an open question what convergence even means under such circumstances.

Acknowledgements The author would like to acknowledge the use of the services provided by Research Computing, University of South Florida.

References

1. T. Belytschko, Y. Krongauz, M. Fleming, D. Organ, W.K. Liu, Smoothing and accelerated computations in the element free galerkin method. J. Comput. Appl. Math. **74**, 111–126 (1996)
2. T. Belytschko, Y. Krongauz, D. Organ, M. Fleming, P. Krysl, Meshless methods: An overview and recent developments. Comput. Methods Appl. Mech. Eng. **139**, 3–47 (1996)
3. D.L. Buchanan, J.H. Gosse, J.A. Wollschlager, A. Ritchey, R. Byron Pipes, Micromechanical enhancement of the macroscopic strain state for advanced composite materials. Compos. Sci. Technol. **69**(11–12), 1974–1978 (2009)
4. J.H. Gosse, S. Christenson, Strain invariant failure criteria for polymers in compostie materials, in *AIAA-2001-1184* (American Institute of Aeronautics and Astronautics, Reston, 2001)
5. W.K. Liu, S. Jun, Y.F. Zhang, Reproducing kernel particle methods. Int. J. Numer. Methods Fluids **20**, 1081–1106 (1995)
6. D.C. Simkins, Multi-scale structural mechanics for advanced aircraft design. J. Nonlinear Syst. Appl. **3**(1), 41–45 (2012)
7. D.C. Simkins, S. Li, Meshfree simulations of thermo-mechanical ductile fracture. Comput. Mech. **38**, 235–249 (2006)
8. T.E. Tay, S.H.N. Tan, V.B.C. Tan, J.H. Gosse, Damage progression by the element-failure method (EFM) and strain invariant failure theory (SIFT). Compos. Sci. Technol. **65**, 935–944 (2005)

Editorial Policy

1. Volumes in the following three categories will be published in LNCSE:

i)   Research monographs
ii) Tutorials
iii) Conference proceedings

Those considering a book which might be suitable for the series are strongly advised to contact the publisher or the series editors at an early stage.

2. Categories i) and ii). Tutorials are lecture notes typically arising via summer schools or similar events, which are used to teach graduate students. These categories will be emphasized by Lecture Notes in Computational Science and Engineering. **Submissions by interdisciplinary teams of authors are encouraged.** The goal is to report new developments – quickly, informally, and in a way that will make them accessible to non-specialists. In the evaluation of submissions timeliness of the work is an important criterion. Texts should be well-rounded, well-written and reasonably self-contained. In most cases the work will contain results of others as well as those of the author(s). In each case the author(s) should provide sufficient motivation, examples, and applications. In this respect, Ph.D. theses will usually be deemed unsuitable for the Lecture Notes series. Proposals for volumes in these categories should be submitted either to one of the series editors or to Springer-Verlag, Heidelberg, and will be refereed. A provisional judgement on the acceptability of a project can be based on partial information about the work: a detailed outline describing the contents of each chapter, the estimated length, a bibliography, and one or two sample chapters – or a first draft. A final decision whether to accept will rest on an evaluation of the completed work which should include

- at least 100 pages of text;
- a table of contents;
- an informative introduction perhaps with some historical remarks which should be accessible to readers unfamiliar with the topic treated;
- a subject index.

3. Category iii). Conference proceedings will be considered for publication provided that they are both of exceptional interest and devoted to a single topic. One (or more) expert participants will act as the scientific editor(s) of the volume. They select the papers which are suitable for inclusion and have them individually refereed as for a journal. Papers not closely related to the central topic are to be excluded. Organizers should contact the Editor for CSE at Springer at the planning stage, see *Addresses* below.

In exceptional cases some other multi-author-volumes may be considered in this category.

4. Only works in English will be considered. For evaluation purposes, manuscripts may be submitted in print or electronic form, in the latter case, preferably as pdf- or zipped ps-files. Authors are requested to use the LaTeX style files available from Springer at http://www.springer.com/authors/book+authors/helpdesk?SGWID=0-1723113-12-971304-0 (Click on Templates → LaTeX → monographs or contributed books).

For categories ii) and iii) we strongly recommend that all contributions in a volume be written in the same LaTeX version, preferably LaTeX2e. Electronic material can be included if appropriate. Please contact the publisher.

Careful preparation of the manuscripts will help keep production time short besides ensuring satisfactory appearance of the finished book in print and online.

5. The following terms and conditions hold. Categories i), ii) and iii):

Authors receive 50 free copies of their book. No royalty is paid.
Volume editors receive a total of 50 free copies of their volume to be shared with authors, but no royalties.

Authors and volume editors are entitled to a discount of 33.3 % on the price of Springer books purchased for their personal use, if ordering directly from Springer.

6. Commitment to publish is made by letter of intent rather than by signing a formal contract. Springer-Verlag secures the copyright for each volume.

Addresses:

Timothy J. Barth
NASA Ames Research Center
NAS Division
Moffett Field, CA 94035, USA
barth@nas.nasa.gov

Michael Griebel
Institut für Numerische Simulation
der Universität Bonn
Wegelerstr. 6
53115 Bonn, Germany
griebel@ins.uni-bonn.de

David E. Keyes
Mathematical and Computer Sciences
and Engineering
King Abdullah University of Science
and Technology
P.O. Box 55455
Jeddah 21534, Saudi Arabia
david.keyes@kaust.edu.sa

and

Department of Applied Physics
and Applied Mathematics
Columbia University
500 W. 120 th Street
New York, NY 10027, USA
kd2112@columbia.edu

Risto M. Nieminen
Department of Applied Physics
Aalto University School of Science
and Technology
00076 Aalto, Finland
risto.nieminen@aalto.fi

Dirk Roose
Department of Computer Science
Katholieke Universiteit Leuven
Celestijnenlaan 200A
3001 Leuven-Heverlee, Belgium
dirk.roose@cs.kuleuven.be

Tamar Schlick
Department of Chemistry
and Courant Institute
of Mathematical Sciences
New York University
251 Mercer Street
New York, NY 10012, USA
schlick@nyu.edu

Editor for Computational Science
and Engineering at Springer:
Martin Peters
Springer-Verlag
Mathematics Editorial IV
Tiergartenstrasse 17
69121 Heidelberg, Germany
martin.peters@springer.com

Lecture Notes in Computational Science and Engineering

48. F. Graziani (ed.), *Computational Methods in Transport*.

49. B. Leimkuhler, C. Chipot, R. Elber, A. Laaksonen, A. Mark, T. Schlick, C. Schütte, R. Skeel (eds.), *New Algorithms for Macromolecular Simulation*.

50. M. Bücker, G. Corliss, P. Hovland, U. Naumann, B. Norris (eds.), *Automatic Differentiation: Applications, Theory, and Implementations*.

51. A.M. Bruaset, A. Tveito (eds.), *Numerical Solution of Partial Differential Equations on Parallel Computers*.

52. K.H. Hoffmann, A. Meyer (eds.), *Parallel Algorithms and Cluster Computing*.

53. H.-J. Bungartz, M. Schäfer (eds.), *Fluid-Structure Interaction*.

54. J. Behrens, *Adaptive Atmospheric Modeling*.

55. O. Widlund, D. Keyes (eds.), *Domain Decomposition Methods in Science and Engineering XVI*.

56. S. Kassinos, C. Langer, G. Iaccarino, P. Moin (eds.), *Complex Effects in Large Eddy Simulations*.

57. M. Griebel, M.A Schweitzer (eds.), *Meshfree Methods for Partial Differential Equations III*.

58. A.N. Gorban, B. Kégl, D.C. Wunsch, A. Zinovyev (eds.), *Principal Manifolds for Data Visualization and Dimension Reduction*.

59. H. Ammari (ed.), *Modeling and Computations in Electromagnetics: A Volume Dedicated to Jean-Claude Nédélec*.

60. U. Langer, M. Discacciati, D. Keyes, O. Widlund, W. Zulehner (eds.), *Domain Decomposition Methods in Science and Engineering XVII*.

61. T. Mathew, *Domain Decomposition Methods for the Numerical Solution of Partial Differential Equations*.

62. F. Graziani (ed.), *Computational Methods in Transport: Verification and Validation*.

63. M. Bebendorf, *Hierarchical Matrices*. A Means to Efficiently Solve Elliptic Boundary Value Problems.

64. C.H. Bischof, H.M. Bücker, P. Hovland, U. Naumann, J. Utke (eds.), *Advances in Automatic Differentiation*.

65. M. Griebel, M.A. Schweitzer (eds.), *Meshfree Methods for Partial Differential Equations IV*.

66. B. Engquist, P. Lötstedt, O. Runborg (eds.), *Multiscale Modeling and Simulation in Science*.

67. I.H. Tuncer, Ü. Gülcat, D.R. Emerson, K. Matsuno (eds.), *Parallel Computational Fluid Dynamics 2007*.

68. S. Yip, T. Diaz de la Rubia (eds.), *Scientific Modeling and Simulations*.

69. A. Hegarty, N. Kopteva, E. O'Riordan, M. Stynes (eds.), *BAIL 2008 – Boundary and Interior Layers*.

70. M. Bercovier, M.J. Gander, R. Kornhuber, O. Widlund (eds.), *Domain Decomposition Methods in Science and Engineering XVIII*.

71. B. Koren, C. Vuik (eds.), *Advanced Computational Methods in Science and Engineering*.

72. M. Peters (ed.), *Computational Fluid Dynamics for Sport Simulation*.

73. H.-J. Bungartz, M. Mehl, M. Schäfer (eds.), *Fluid Structure Interaction II - Modelling, Simulation, Optimization.*

74. D. Tromeur-Dervout, G. Brenner, D.R. Emerson, J. Erhel (eds.), *Parallel Computational Fluid Dynamics 2008.*

75. A.N. Gorban, D. Roose (eds.), *Coping with Complexity: Model Reduction and Data Analysis.*

76. J.S. Hesthaven, E.M. Rønquist (eds.), *Spectral and High Order Methods for Partial Differential Equations.*

77. M. Holtz, *Sparse Grid Quadrature in High Dimensions with Applications in Finance and Insurance.*

78. Y. Huang, R. Kornhuber, O.Widlund, J. Xu (eds.), *Domain Decomposition Methods in Science and Engineering XIX.*

79. M. Griebel, M.A. Schweitzer (eds.), *Meshfree Methods for Partial Differential Equations V.*

80. P.H. Lauritzen, C. Jablonowski, M.A. Taylor, R.D. Nair (eds.), *Numerical Techniques for Global Atmospheric Models.*

81. C. Clavero, J.L. Gracia, F.J. Lisbona (eds.), *BAIL 2010 – Boundary and Interior Layers, Computational and Asymptotic Methods.*

82. B. Engquist, O. Runborg, Y.R. Tsai (eds.), *Numerical Analysis and Multiscale Computations.*

83. I.G. Graham, T.Y. Hou, O. Lakkis, R. Scheichl (eds.), *Numerical Analysis of Multiscale Problems.*

84. A. Logg, K.-A. Mardal, G. Wells (eds.), *Automated Solution of Differential Equations by the Finite Element Method.*

85. J. Blowey, M. Jensen (eds.), *Frontiers in Numerical Analysis - Durham 2010.*

86. O. Kolditz, U.-J. Gorke, H. Shao, W. Wang (eds.), *Thermo-Hydro-Mechanical-Chemical Processes in Fractured Porous Media - Benchmarks and Examples.*

87. S. Forth, P. Hovland, E. Phipps, J. Utke, A. Walther (eds.), *Recent Advances in Algorithmic Differentiation.*

88. J. Garcke, M. Griebel (eds.), *Sparse Grids and Applications.*

89. M. Griebel, M. A. Schweitzer (eds.), *Meshfree Methods for Partial Differential Equations VI.*

For further information on these books please have a look at our mathematics catalogue at the following URL: www.springer.com/series/3527

Monographs in Computational Science and Engineering

1. J. Sundnes, G.T. Lines, X. Cai, B.F. Nielsen, K.-A. Mardal, A. Tveito, *Computing the Electrical Activity in the Heart.*

For further information on this book, please have a look at our mathematics catalogue at the following URL: www.springer.com/series/7417

Texts in Computational Science and Engineering

1. H. P. Langtangen, *Computational Partial Differential Equations.* Numerical Methods and Diffpack Programming. 2nd Edition

2. A. Quarteroni, F. Saleri, P. Gervasio, *Scientific Computing with MATLAB and Octave.* 3rd Edition

3. H. P. Langtangen, *Python Scripting for Computational Science.* 3rd Edition

4. H. Gardner, G. Manduchi, *Design Patterns for e-Science.*

5. M. Griebel, S. Knapek, G. Zumbusch, *Numerical Simulation in Molecular Dynamics.*

6. H. P. Langtangen, *A Primer on Scientific Programming with Python.* 3rd Edition

7. A. Tveito, H. P. Langtangen, B. F. Nielsen, X. Cai, *Elements of Scientific Computing.*

8. B. Gustafsson, *Fundamentals of Scientific Computing.*

9. M. Bader, *Space-Filling Curves.*

10. M. Larson, F. Bengzon, *The Finite Element Method: Theory, Implementation and Applications.*

For further information on these books please have a look at our mathematics catalogue at the following URL: www.springer.com/series/5151